Society, Environment and Human Security in the Arctic Barents Region

The Arctic-Barents Region is facing numerous pressures from a variety of sources, including the effect of environmental changes and extractive industrial developments. The threats arising out of these pressures result in human security challenges.

This book analyses the formation, and promotion, of societal security within the context of the Arctic-Barents Region. It applies the human security framework, which has increasingly gained currency at the UN level since 1994 (UNDP), as a tool to provide answers to many questions that face the Barents population today. The study explores human security dimensions such as environmental security, economic security, health, food, water, energy, communities, political security and digital security in order to assess the current challenges that the Barents population experiences today or may encounter in the future. In doing so, the book develops a comprehensive analysis of vulnerabilities, challenges and needs in the Barents Region and provides recommendations for new strategies to tackle insecurity and improve the wellbeing of both indigenous and local communities.

This book will be a valuable tool for academics, policy-makers and students interested in environmental and human security, sustainable development, environmental studies and the Arctic and Barents Region in particular.

Kamrul Hossain is a Research Professor and Director of the Northern Institute for Environmental and Minority Law, University of Lapland.

Dorothée Cambou is a post-doctoral researcher at the University of Helsinki.

Routledge Explorations in Environmental Studies

For a full list of titles in this series, please visit https://www.routledge.com/Routledge-Explorations-in-Environmental-Studies/book-series/REES

The Governance of Green Urban Spaces in the EU
Social innovation and civil society
Judith Schicklinski

NGO Discourses in the Debate on Genetically Modified Crops
Ksenia Gerasimova

Sustainability in the Gulf
Challenges and Opportunities
Edited by Elie Azar and Mohamed Abdelraouf

Environmental Human Rights
A Political Theory Perspective
Edited by Markku Oksanen, Ashley Dodsworth and Selina O'Doherty

African Philosophy and Environmental Conservation
Jonathan O. Chimakonam

Domestic Environmental Labour
An Eco-feminist Perspective on Making Homes Greener
Carol Farbotko

Stranded Assets and the Environment
Risk, resilience and opportunity
Edited by Ben Caldecott

Society, Environment and Human Security in the Arctic Barents Region
Edited by Kamrul Hossain and Dorothée Cambou

Environmental Performance Auditing in the Public Sector
Enabling Sustainable Development
Awadhesh Prasad

Society, Environment and Human Security in the Arctic Barents Region

Edited by
Kamrul Hossain and Dorothée Cambou

Routledge
Taylor & Francis Group
LONDON AND NEW YORK

earthscan
from Routledge

First published 2018
by Routledge
2 Park Square, Milton Park, Abingdon, Oxon OX14 4RN

and by Routledge
711 Third Avenue, New York, NY 10017

Routledge is an imprint of the Taylor & Francis Group, an informa business

British Library Cataloguing in Publication Data
A catalogue record for this book is available from the British Library

Library of Congress Cataloging in Publication Data
Names: Hossain, Kamrul, editor. | Cambou, Dorothâee, editor.
Title: Society, environment and human security in the Arctic Barents region / edited by Kamrul Hossain and Dorothâee Cambou.
Description: Abingdon, Oxon ; New York, NY : Routledge, 2018. | Series: Routledge explorations in environmental studies | Includes bibliographical references and index.
Identifiers: LCCN 2017061084 (print) | LCCN 2018015027 (ebook) | ISBN 9781351171243 (eBook) | ISBN 9780815399841 (hardback)
Subjects: LCSH: Barents Sea Region--Environmental conditions. | Barents Sea Region--Social conditions. | Human security--Barents Sea Region.
Classification: LCC GE160.B33 (ebook) | LCC GE160.B33 S64 2018 (print) | DDC 304.20948--dc23
LC record available at https://lccn.loc.gov/2017061084

ISBN: 978-0-8153-9984-1 (hbk)
ISBN: 978-0-367-85581-9 (pbk)
ISBN: 978-1-351-17124-3 (ebk)

Typeset in Times New Roman
by Taylor & Francis Books

Contents

Illustrations

Figure

Tables

Contributors

Editors

Kamrul Hossain is a Research Professor and the Director of the Northern Institute for Environmental and Minority Law (NIEM) at the Arctic Centre in the University of Lapland. He has previously served the Faculty of Law of the University of Lapland as the (acting) Professor of Public International Law. He is also an Adjunct Professor of International Law at the University of Lapland. At present he leads two significant research projects – one funded by the Nordforsk having partners from Norway, Sweden and the UK, and the other funded by the Academy of Finland – highlighting various dimensions of human security as it is applicable to the Arctic and the European High North. Dr Hossain has extensively published scholarly articles in the field of international law as well as in law and human rights that apply to the Arctic. He served as the Special Editor of the *Yearbook of Polar Law* published in 2011. He regularly teaches at the University of Lapland and periodically at other foreign universities. He completed visiting fellowships at a number of foreign universities including the University of Toronto, Scott Polar Research Institute of the University of Cambridge, and the Muroran Institute of Technology in Japan. He has received a number of prestigious awards including SYLFF Fellowship at the University of Helsinki. In 2012, he was awarded a Faculty Research Fellowship by the Canadian Department of Foreign Affairs and International Trade (DFAIT).

Dorothée Cambou is a post-doctoral researcher at the University of Helsinki. Formerly, she was conducting post-doctoral research at the Northern Institute for Environmental and Minority Law (NIEM) at the Arctic Centre in the University of Lapland. Currently, the main focus of her research lies in human rights law in particular concerning the rights of local communities and indigenous peoples. She has also developed an expertise in the field of sustainable development and Arctic governance and cultivates a solid interest for issues concerning business and human rights. She is now leading an independent research project on energy justice and the rights of indigenous peoples. Dr Cambou also holds a PhD of laws from the Vrije University Brussel, where she defended a thesis on the legal significance of the right of indigenous peoples to self-determination and its implications for the Sámi people.

Assistant editors

Joëlle Klein is a Research Assistant at the University of Lapland's Arctic Centre in the Northern Institute for Environmental and Minority Law (NIEM). She is a recent graduate with a Master's in Human Rights and Democratisation from the European Inter-University Centre and the Institute of Human Rights at Åbo Akademi. Her Master's thesis focused on indigenous rights in environmental justice campaigns, specifically looking at the role of colonialism and applicability of human security in the context of anti-sealing campaigns and Inuit seal hunting. At the Arctic Centre, she has provided editorial assistance and supported research on human security in the Barents, Sámi rights to self-determination and the rights of minorities within minorities. She has also participated in the HusArctic project, presented research at the 10th Polar Law Symposium and was recently appointed as a special editor for the 10th volume of the *Yearbook of Polar Law*. Her research interests include human rights, specifically indigenous and minority rights, human security, environmental activism and natural resource governance in the arctic, as well as the role of gender and intergenerational perspectives in contemporary social change.

Karolina Siroka is a research assistant at the Northern Institute for Environmental and Minority Law (NIEM), at the Arctic Centre, University of Lapland. She holds a master's in law from University of Wroclaw, where she defended a thesis in the Department of Material Criminal Law. Before that, she was working in a law firm in Poland, where she focused on civil and criminal law. Currently Karolina is working on the rights of indigenous peoples, regarding land rights as well as the protection of cultural heritage.

Authors

Giuseppe Amatulli is a PhD Candidate in the programme 'Northern Cultures and Sustainable Natural Resources Politics', Faculty of Social Sciences, University of Lapland. Giuseppe is based at the Arctic Centre, where he works also as a researcher, and is part of the Sustainable Development Research Group. Before arriving in Rovaniemi, Giuseppe lived and studied in different countries. He gained his bachelor's and master's degrees in Political Science and International Relations from the University of Trieste (Italy) and after that he moved to live in Wales (Cardiff) for a while. In the meanwhile he was admitted in the EMA (European Master's Degree in Human Rights) programme and he studied in Venice and Turku, where he wrote his master's thesis at the Institute for Human Rights of Åbo Akademi University. His thesis, entitled 'The legal position of the Sámi in the exploitation of mineral resources in Finland, Norway and Sweden', was defended in September 2014 and was awarded with publication in February 2015. After the EMA programme, Giuseppe worked in Luxembourg, in a Development NGO, carrying out research regarding microfinance institutions and enhancement of human rights. After such experience he wanted to gain field experience and for this reason he decided to work in a refugee camp in Italy for six months. In December 2016 Giuseppe was admitted in the PhD programme of the University of Lapland and his research topic is related with the inclusion of Sámi people in decision-making process, with a particular focus on local communities and their protection.

Patrick Ciaschi is a PhD candidate in the Politics department at the New School for Social Research (NSSR). His dissertation focuses on experiences of displacement and forced mobility and their relationship to shifting property regimes within re-established 'illiberal' post-socialist states. Since 2014, he has been specifically working within marginalised Roma communities in current day Hungary to comprehend both everyday illiberal orders and cultures of removal that pervade localised power structures, concern personal, discursive and subjective understandings of belonging, security and identity. His work draws from interdisciplinary subfields of critical International Relations and Global Politics including citizenship studies, critical security studies and migration studies. Currently, he is a Visiting Scholar at the University of Debrecen and has been an active member of, and research assistant to, the Zolberg Institute on Migration and Mobility (ZIMM) housed at the New School.

Shaun Cormier was a contract researcher with the Arctic Centre, where he was working on Food Security related research with a focus on the Barents Region. Shaun's relation to the topic of food security was born from growing up on a trap line and living off the land in northern Canada in the earlier years of his life. His interest in Food Security and the Barents Region has since developed after publishing his Master's thesis on 'Enhancing Indigenous Food Security in the Arctic: Through Law, Policy, and the Arctic Council' one year ago. He admires those who continue to live in sustainable ways and work to make the world more food secure.

Lassi Heninen is a Professor of Arctic Politics at the Faculty of Social Sciences, University of Lapland (Finland), and Visiting Professor at University of Akureyri (Iceland). His research fields include IR, Geopolitics, Security Studies, Environmental Politics, Northern/Arctic Studies and Political History. Prof. Heininen teaches and lectures regularly abroad and supervises MA and PhD students from Europe, Russia and North America. Among his other academic positions are Senior Research Scholar (associate) at

IIASA, Adjunct Professor at Trent University (Ontario, Canada) and Director of International Summer School in Karelia at Petrozavodsk State University (Russia). He is also Head of (UArctic-NRF) Thematic Network on Geopolitics and Security Studies and the Editor of the *Arctic Yearbook*, a peer-reviewed online international publication (at www.arcticyearbook.com). Among his recent scientific publications on the Arctic and Barents Sea area are 'The Arctic, Baltic and North Atlantic Cooperative Regions in the "Wider Northern Europe"' in *Journal of Baltic Studies* (2016); *Future Security of the Global Arctic: State Policy, Economic Security and Climate* by Palgrave Pivot (2016); 'The Barents Region in the State Interests and International Politics' in *Barents Journal* (2009).

Hanna Lempinen is a Post-Doctoral Researcher at the Aleksanteri Institute at the University of Helsinki, Finland, and a Visiting Researcher at the Arctic Centre, University of Lapland. She obtained her PhD from the University of Lapland in 2017 with a doctoral dissertation focusing on the social dimension of the Arctic energyscape. While her background is formally in the field of International Relations, in her work she is combining perspectives from environmental sociology, sustainability studies and welfare research in developing new approaches to energy as a societal concern. Alongside her research Lempinen has worked as a university teacher and a freelance journalist reporting on energy-related affairs.

Sarah Mackie is a PhD candidate at Newcastle University writing a doctoral thesis on comparative environmental law in the Arctic. Sarah has held positions as a Visiting Researcher at the Arctic Centre at the University of Lapland, the University of Greenland and Harvard Law School. Sarah completed an undergraduate law degree at Cambridge University and an LL.M. in Environmental Regulation at Newcastle University. She has worked at an environmental law firm in Anchorage, Alaska, and as Judicial Assistant to the Lord Chief Justice of England and Wales.

Susanna Pääkkölä is a research scientist with a M.Sc. focused in Faculty of Science, Department of Biology (Physiology and Genetics) from the University of Oulu, Finland. She works as a research scientist at the Arctic Centre, University of Lapland, Rovaniemi, Finland, and holds a position as a Doctoral Candidate in Biomedicine at the University of Oulu, Finland. Susanna has working experience in the research industry and education sector. Currently she works on human physiology and thermoregulation for her PhD and works as a researcher at the Arctic Center.

Anna Petrétei is working as a researcher at the Northern Institute for Environmental and Minority Law (NIEM) and is a PhD candidate at the University of Lapland, Faculty of Law. She graduated as a lawyer at the University of Pécs, Hungary. She is carrying out research on the rights of indigenous peoples. She is currently involved in a project focusing on human security in the Arctic Region (Human Security as a promotional tool for societal security in the Arctic: Addressing Multiple Vulnerability to its Population with Specific Reference to the Barents Region), where she examines different aspects of societal and indigenous security. She has published several articles on indigenous rights related to mining activities. Besides studying Sami rights and traditional livelihoods, she continues her research by expanding her approach to other indigenous peoples. Her research interests are minority law, indigenous rights, traditional livelihoods and mining law from the viewpoint of indigenous peoples.

Tahnee Prior is a 2015 Trudeau Scholar, Vanier Canada Graduate Scholar and PhD candidate in global governance at the University of Waterloo. Her dissertation seeks to understand what a legalised international governance structure that could account for the complexity and rapid change of the Arctic could look like and how we could get there from here. Tahnee is a team member of an Academy of Finland research project on Human Security as a Promotional Tool for Societal Security in the Arctic, a Visiting Researcher with the Arctic Futures Initiative at the International Institute for Applied Systems Analysis and the lead of Plan A, a storytelling platform on gender and the Arctic. Tahnee holds a BA in international relations from Franklin University Switzerland and an MA in global governance, specialising in multilateral institutions and diplomacy, from the Balsillie School of International Affairs.

Dele Raheem is a Senior Researcher at the Northern Institute for Environmental and Minority Law (NIEM)/Arctic Centre, University of Lapland. He obtained his doctoral degree in Food Sciences from the Department of Food and Environmental Sciences at the University of Helsinki, Finland. He has conducted nutritional and microbiological research on different indigenous food crops, including their processing methods. His current research interest is on crosscutting issues related to food and human security especially in the Arctic-Barents Region. He is a member of the Food Climate Research Network based at the University of Oxford, UK.

Mirva Salminen conducts her doctoral study on digitalisation and cybersecurity in the European High North at the Arctic Centre, University of Lapland. Previously, she has carried out research for several organisations in both public and private sectors, including universities, research institutes and corporations, publishing primarily on cybersecurity and security commercialisation. She is also a recognised public speaker on cybersecurity. Salminen holds MSSc degree in International Politics and BSSc in Political Science, both from the University of Tampere. Parts of her degrees she has completed in the US and in the UK.

Antonia Sohns is a PhD student at McGill University where she is researching water vulnerability in Arctic households under the supervision of Dr James Ford. Before her studies at McGill, Antonia worked as a water and energy analyst and consultant for the World Bank's Thirsty Energy initiative in

Washington, DC. Antonia has an MSc in Water Science, Policy and Management from the University of Oxford, where her thesis focused on the water-energy nexus and the water requirements of hydraulic fracturing operations in Pinedale, Wyoming. She also holds a BS in Earth Systems, Oceans track from Stanford University.

Preface

In academic literature, the Arctic Barents Region is relatively underexplored. While Arctic challenges addressed in both academic and policy documents also include regional features prevailing in the Barents Region, the latter is somewhat distinct compared to general Arctic features. The Barents Region is generally composed of both the European high Arctic and its sub-Arctic regions, including northern parts of Finland, Norway and Sweden (the Nordic Barents), and the larger Northwestern Russia (Russian Barents). In terms of socio-cultural, economic, political and infrastructural development, the Nordic Barents has seen greater advancement than the Russian Barents. However, the entire region faces common environmental challenges exacerbated by industrial processes and climate change. Interest for developing, new projects such as mining and other extractive industrial activities also continue to increase across both parts of the Barents. Local and indigenous communities are managing challenges to the loss of traditional subsistence activities in response to increasing human activity. Navigation and offshore oil and gas resource extraction are becoming more open for international usage in the marine regions of the Barents. As a byproduct of industrial development and globalisation, external actors and forces are entering the region. On the one hand, these developments bring environmental challenges, but on the other hand, they also offer economic incentives for the region to prosper. However, the benefits arising out of these new economic developments and human activities apparently offer less opportunity to those of the communities living in remote areas. As a result unequal benefit sharing from development activities in the Barents economy is resulting in tensions amongst the various groups living in the region.

In this book, our theoretical analysis is structured around the concept of societal security, which is a concept yet to concretely develop. However, given the uniqueness in regard to the shared geographical and socio-environmental features as well as shared consequence of the transformation of the region, we point the Barents Arctic Region as a transnational society – the society which is threatened by a number of stressors. We are convinced that many of the threats prevailing in the region are synonymous to human security challenges. We view that the promotion of human security will eventually promote societal security in the region, and provide a number of recommendations or suggestions to further explore based on the research conducted by the individual chapter authors.

This book is the product of a research project entitled: 'Human Security as a promotional tool for societal security in the Arctic: Addressing Multiple Vulnerability to its Population with Specific Reference to the Barents Region (HuSActic)' funded by the Academy of Finland. We gratefully acknowledge this financial contribution in supporting our ability to carry out this research. In the process of preparing this manuscript, we also relied on a number of persons who both directly and indirectly supported us including a number of peer reviewers, and Timo Koivurova Director of the Arctic Center at the University of Lapland, whose comments and suggestions throughout the process significantly contributed to our research. We also greatly acknowledge the research assistance and support provided by Joëlle Klein and Karoliną Sikora. Finally, we are grateful to all the chapter authors for their patience and hard work during the writing and revision processes.

Kamrul Hossain and Dorothée Cambou
30 November 2017

General introduction and the structure of the book

Kamrul Hossain and Dorothée Cambou

The Barents Region, compared to other parts of the circumpolar Arctic, is a relatively densely populated region. Among others, the original population of the region consists of a number of indigenous groups including the Sámi of Fennoscandia, who transnationally inhabit all four of the Barents countries. Also, in comparison to other Arctic regions, the Barents is more developed in terms of both physical infrastructure and social networks. The region's autonomous identity has been developing since the beginning of the 1990s after the breakdown of the Soviet Union and the end of the Cold War. During the Cold War, the region was the object of a reified conceptualisation of security that was the purview of national interests and high politics. At the end of the Cold War, high politics were replaced by a new agenda, forming the regions' new identity. The recognition of particular concerns, such as the threat to its pristine environment affecting the region at large, was developing in the 1980s, and slowly became a broader priority. In 1987, the former Soviet leader Mikhail Gorbachev's famous speech in Murmansk was a milestone in raising awareness for the region's emerging issues – environmental protection. He declared the whole of the Arctic Region to be considered as a 'zone of peace' (Gorbachev 1987), with a desire to create a nuclear-free Northern Europe. This development represented a new dynamic in the context of regional affairs. It was not an isolated event, as the simultaneous development of security as a regional concept was also growing. From the beginning of the 1990s, a number of schools of thought in security studies developed further conceptualisations of security, by analysing ontological aspects of security as well as by broadening and deepening the concept. Security developed a new dimension that challenged its use as solely a mechanism to deal with state sovereignty and inter-state matters.

The conceptualisation of security as only protecting states' border, territory and national interests was replaced by other agenda, namely the protection of environment, and by extension the promotion of human wellbeing. The Barents Region is an interesting example of the integration of multiple challenges, mostly deriving from risks associated with environmental changes, which can be classified in the contemporary discourse of human security. The sustenance of the whole of the region in its unique form depends on the protection and promotion of its natural environment. Hence, the first formal step in an effort to develop an inter-state cooperation process was the adoption of the so-called Rovaniemi Declaration adopted in 1991 which included protections for the Arctic environment. By virtue of this process, the Arctic Environmental Protection Strategy (AEPS) has been endorsed by the eight circumpolar countries (Koivurova 2005, 208; Nuttall 2005, 116). Two years later in 1993 another forum was established – the Barents Euro-Arctic Council – in order to promote further systematic and coherent cooperation process within this European sub-region of the Arctic, namely the Barents (Bailes and Ólafsson 2017; Olsson et al. 2016). Today, the AEPS has been integrated into the Arctic Council, which was established in 1996 as a high-level intergovernmental forum of the eight arctic states (Nord 2015; Koivurova 2005; Young 1996). These institutional efforts continue to progressively safeguard and promote the environmental, economic and social-cultural sustainability of the region.

The term 'sustainability' is rather ambiguous, unless precisely conceptualised in a given context. It can be a source of conflict in considering the interests of different stakeholders and actors. For example, talking about environmental sustainability can often conflict with economic sustainability, and vice versa. At the same time, socio-cultural sustainability is dependent on both environmental and economic sustainability. Therefore, contextualising sustainability in the Barents Region requires contextualising specific challenges, including environmental changes facing the region, that have socio-cultural and economic effects on the region's inhabitants. In addition, promoting mechanisms to deal with these challenges in a way that preserves a continuity of societal identity supports a more resilient form of sustainability. Throughout this book, we use the language of human security, which seeks to protect 'the vital core of all human lives' (CHS 2003, 4), and ensure freedom and human fulfilment, and promote human wellbeing which provides a resilient sustainability through preserving the continuity of authentic societal existence.

The concept of human security, as defined in the Human Development Report of the United Nations Development Program (UNDP 1994) is linked to a number of issues that influence societal sustainability, and thereby promote societal security (see Chapter 1.1). Defining the concept 'societal security' is complex given that the formation of a society is linked to both structural and functional conditions at different levels. When the ability of a society to 'persist in its essential character under changing conditions and possible or actual threats' is challenged, then the society's existence is threatened (Waever et al. 1993, 23). Within the Barents Region, the application of a relevant human security agenda in policy processes would promote societal security in the regional context. Despite differences between the referent objects in human security and societal security, the concepts can still be interlinked in the Barents context. Threats to a society's collective identity include 'the suppression of its expression, to interference with its ability to reproduce' (Waever et al. 1993, 21). In the Barents, societal insecurity exists in not only threats to cultural identity, but also threats to health, economy, environment, etc. These threats are not distinct – they are human security concerns threatening the 'ability to reproduce' for its population in the region.

The concept of human security has gained significance in the Arctic-Barents Region in recent years, as it offers a different angle to understanding the complex threats facing its communities. The link between human security, climate change and the environment has been significantly endorsed in a number of available literatures and especially in Mark Nuttall's contribution relating to the Arctic Region (e.g. Gjørv et al. 2013; Nuttalls 2013); human security's link to societal integrity in the regional context so far remains abstract. Additionally, a substantial focus on addressing the Barents Region in specific regional terms has not been properly explored in this context. Given that states are no longer the only subject of security, and that the focus of security discourse has shifted to the protection of individuals and communities from a wide range of interconnected threats, the conceptualisation of security in the Barents Region must be inclusive. Similarly, the institutional arrangements referred to above play a diffuse role in the governance of the region. The governance approach includes the participation of new actors demonstrating that the governance of the region must be developed on shared understandings between various stakeholders at varied levels. Thus, taking a policy-oriented approach, we explore the human security challenges faced by the Barents population in order to address the new and particular causes of insecurities affecting the Barents society at large.

The Arctic-Barents Region has unique features, including its pristine natural environment, the presence of distinct communities including a number of indigenous communities with distinct livelihood practices, cultures and subsistence activities. The communities in the Arctic-Barents Region share norms, values and institutions connected to the region's natural environment, which provide them with an incentive for a cohesive society. However, changes occurring in the region affect such cohesion. While some changes bring new opportunities for the region, others adversely affect individuals and communities as the socio-environmental and cultural integrity held by its population is threatened.

The overall progression influences the continuity of societal identity in its own form. We have conceived that the idea of formation, of a society or its continuity is dependent on particular common issues at various levels (see Chapter 1.1), and defined by historical, geographical, political and environmental relations prevailing amongst the particular society in a given regional context. Thus, our perception of the homogeneity of Barents society formed around particular common challenges. For example, threats posed by climate change and an increase in industrial activities are common across the region and impact all communities, although their manifestation is context specific. Methodologically, we use the concept of human security as a framework to improve societal security within the Barents Region. In accordance with the human security approach introduced in the 1994 UNDP Human Development

Report, we focus on economic, health, environmental, food, personal, community and political security as means to assess current challenges in the Barents Region. In addition, we have included water security, energy security and digital security in our analysis in order to offer a more comprehensive assessment. While evaluating human security can be challenging, each security dimension will include a unique definition of assessment variables, to provide the reader with context-specific understandings of its role in the Barents Region. The resulting analysis will constitute a critical starting point for policy-making focused on the development of the Barents Region and the wellbeing of its population.

The book is organised into two main sections. Part 1 is composed of two chapters that set out the theoretical and contextual background for subsequent assessments. In the first chapter, Kamrul Hossain provides an analysis of the theoretical framework used in our assessment and defines the main concept of societal security, its link to human security and relevance in the Barents Region. Then, Dorothée Cambou and Lassi Heninen provide contextual background with a descriptive analysis of the Barents Region, focusing on the institutional development of the Barents cooperation. Part 2 comprises the subsequent security assessments, and is composed of ten chapters. Each chapter focuses on one of the ten human security dimensions: environmental, economic, health, food, water, energy, personal, community, political and digital security. We invited a diverse group of scholars to discuss each dimension of human security. Although all contributors were asked to structure their arguments identically – Definition, Assessment, Conclusion – they were also invited to openly address other specific issues that they considered relevant for the analysis. The final section of the book includes a short summary of the principal recommendations that could be useful for policy development in the region.

Human security is multifaceted, and threats to security are interconnected. In the Barents Region, environmental threats constitute a major challenge to the immediate and long-term wellbeing of the population, and considerably affect all other human security dimensions. Environmental changes can for instance undermine the economic prosperity of the region, which relies on primary sectors such as forestry, extractive industries and agriculture. In turn, environmental threats also have implications for the livelihoods and health of existing communities. Thus, environmental security is an overreaching dimension of human security in the Barents Region, which also feeds into all other dimension of human security. Therefore, the first chapter of Part 2 focuses on the assessment of environmental security and is followed by two chapters on economic and health security.

More precisely, in Chapter 2.1, Sarah Mackie analyses the specific challenges that affect the environment of the Barents Region. After defining environmental security and describing the different habitats found across the region, the chapter considers three major threats to the environment: climate change, industrial pollution and nuclear contamination. In the following chapter, Anna Petrétei and Dorothée Cambou present economic security as it applies to the Barents Region. They expand the concept of economy security to go beyond economic growth or the market sphere, and to assess the capacities of peoples to sustain adequate standard of living in accordance with their own needs, cultural standards and the environment. In their analysis, they examine factors in demographics, migration, education, transportation and existing conflicts with extractive industries as elements of economic security. Chapter 2.3 examines key health security issues in the Barents population. In this chapter, Susanna Pääkkölä and Dorothée Cambou describe the impact of environmental challenges on the health situation of the population, including its cold climate, transboundary pollution and climate change. The analysis also includes references to the specific epidemiological situation of the population, and the existence of different trends in Russia and the Nordic countries. The chapter ends with considerations on the link between mental health and the wellbeing of indigenous communities.

The next three chapters focus on the food-water-energy nexus. While it is clear that improving food, water and energy security requires an integrated approach to the governance of all three resources, the analysis is divided into three chapters in order for the authors to provide recommendations based on their specific research expertise. In Chapter 2.4 on 'Food security in the Barents Region', Dele Raheem and Shaun Cormier highlight for instance the important contribution of traditional and local foods to the food security of indigenous and non-indigenous people in the Barents Region. They also briefly underline how globalisation affects food systems to challenge food security in the Barents Region. Chapter 2.5, written by Antonia Sohns, explores how Arctic freshwater systems are threatened by concomitant impacts of aging and insufficient infrastructure, impaired freshwater resources, ecosystem degradation, industrial development, climate change and demographic shifts. Chapter 2.6 by Hanna Lempinen and Dorothée Cambou focuses on societal dimensions of energy security in the region. Lempinen and Cambou draw attention to the contextual nature of energy security and through the dimensions of 1) availability, 2) affordability, 3) environmental aspects, 4) acceptability and 5) the diversity of energy sources as well as socio-cultural diversity inhabitants, including indigenous peoples, in the Barents Region.

A final but central aspect of the assessment focuses on the protection of individuals and their cultures and political rights, through the study of personal, community and political security. Chapter 2.7 on personal security is written by Tahnee Prior and Patrick Ciashi and focuses on three particular dimensions of personal (in)security: suicide, domestic violence, migration and mobility. In exploring some of these areas, the authors focus on two groups in particular: women and indigenous peoples. They do so to pinpoint the evolving tensions and conflicts related to personal insecurity as they arise in the Arctic and the Barents Region, more specifically. Ultimately, their assessment also highlights areas where current work remains insufficient or imminent. Chapter 2.8 focuses on community security. Although personal security is vital to human security, most peoples derive security from their membership in a group (UNDP

1994). In this regard, Giuseppe Amatulli and Joelle Klein address community security in the context of the Barents Region, with a particular focus on indigenous peoples living in the area. In their analysis, the authors stress emphasis on the promotion and preservation of indigenous languages as a central aspect of identity, and argue that access and rights to traditional lands and natural resource management pose a serious risk to the sustainability of traditional lifestyles. Chapter 2.9 focus more broadly on the issue of political security in the Barents Region. In this chapter, Dorothée Cambou argues that although traditionally loosely defined, political security gains currency in the framework of human rights, as it includes both the individual rights of the citizens and the collective rights of indigenous peoples, through the defence of democracy, peace and stability. The final assessment chapter by Mirva Salminen addresses the inter-linkage between digitalisation and human security. More precisely, Mirva Salminen examines digital security under six themes: (1) access to digital networks, (2) availability of digital services, (3) access to (relevant) information, (4) digital awareness and skills, (5) digital inclusion policies and protection of human rights, and (6) state of cybercrime and digital abuse. Her chapter places the wellbeing of individuals and communities in the Barents Region at the centre of digital security hence broadening the conceptualisations of, for instance, information security and cybersecurity, which traditionally focus on the protection of information or national security. In addition, the chapter serves as a reminder of the importance of empowerment alongside protection for the constitution of a truly comprehensive digital security framework.

While each contribution represents several distinct perspectives on human security issues in the Barents area, the overall assessment provides a general understanding of key challenges faced by the population. In this regard, we believe the assessment will not only serve as a basis for knowledge development, future scientific inquiry and policy development, but also contribute to improving overall human security in the Barents Region.

References

Bailes, A.J.K. and Ólafsson, K.Þ. 2017. The EU crossing Arctic frontiers: the Barents Euro-Arctic Council, Northern Dimension, and EU-West Nordic relations. In *The European Union and the Arctic*, edited by N. Liu, E.A. Kirk and T. Henriksen. Leiden, Boston: Brill.

Commission on Human Security (CHS). 2003. *Human Security Now*. New York: CHS.

Gjørv, G.H., Bazely, D., Goloviznina, M. and Tanentzap, A. (Eds.). 2013. *Environmental and Human Security in the Arctic*. London: Routledge.

Gorbachev, M. 1987. Speech at the ceremonial meeting on the occasion of the presentation of the order of Lenin and the gold star of the city of Murmansk.

Koivurova, T. 2005. Environmental protection in the Arctic and Antarctic: can the polar regimes learn from each other. *International Journal of Legal Information: The Official Publication of the International Association of Law Libraries* 33(2): 204–218.

Nord, D.C. 2015. *The Arctic Council: Governance within the Far North*. London: Routledge.

Nuttall, M. 2005. *Encyclopedia of the Arctic*. London: Routledge.

Nuttall, M. 2013. Climate change and human security in the Arctic. In *Handbook on Climate Change and Human Security*, edited by M.R. Redclift and M.Grasso. Cheltenham: Edward Elgar Publishing.

Olsson, M.O., Backman, F., Golubev, A., Norlin, B., Ohlsson, L. and Elenius, L. 2016. *Encyclopedia of the Barents Region*. Oslo: A-M. Pax.

United Nations Development Programme. 1994. *UN Development Report*. New York: Oxford University Press.

Waever, O., Buzan, B., Kelstrup, M. and Lemaitre, P. (Eds.) 1993. *Identity, Migration and the New Security Agenda in Europe*. New York: Palgrave Macmillan.

Young, O.R. 1996. *The Arctic Council: Marking a New Era in International Relations*. New York: Twentieth Century Fund Report.

Part I

Theory and context

1.1 The question of societal security in the Arctic

Kamrul Hossain

1.1 Society and a transnationally organised community

The formation of a 'society' is complex, given that there is no set definition for it. According to the Cambridge English Dictionary (Cambridge Dictionary, n.d.), society is:

> a large group of people who live together in an organised way, making decisions about how to do things and share the work that needs to be done. All the people in a country, or in several similar countries, can be referred to as a society.

This definition suggests that a society can be formed both within and beyond the domain of a nation state. Generally, individuals are the core of a society, living in a particular geographical setting and having common issues to share. They interact with each other and bind themselves in a functional relationship with each other by virtue of a number of mutually recognised identical norms broadly known as culture. The practising of culture provides a sense of cohesion by which members within the community identify each other as part of the same group. However, this cohesion may not always have to be homogenous in a strict sense, as there are smaller groups within bigger groups, thus forming both smaller societies and bigger societies.

The construction of a social group is based on particular likeness amongst its members. Persons with similar interests join together and build their smaller societies within a bigger society. For example, individuals from the same profession find a likeness in their belonging to that particular profession. Yet, there is a shared understanding that an individual also belongs to a broader identity or broader collective identity at large, which also distinguishes them from other outside groups (Ross 2009, 58). Individuals find their affinity to a group through the particular surroundings that form its culture. It is 'the set of meanings owned by a given culture that sets it apart from other cultures' (Keillor et al. 1996, 58). The factors that distinguish a group from 'others' are largely based on representations of a common historical heritage, symbols, customs and traditions, which offer the in-group community a basic homogeneity.

Such a homogenous society or community is built around the common features prevailing in a particular geographical location, which are the result of a so-called 'we feeling'. Such features may have international, inter-regional, trans-national, national or sub-national dimensions. Internationally, for example, the 'we feeling' can be built around the common interests of humankind, such as protecting the planet from the global effects of climate change. We unite ourselves together as an international community to fight against the threats that an increased concentration of greenhouse gasses pose to the planet. At the regional level, for example, the European Union (EU) – a supra-national body – finds affinity around the common values of democracy, human rights, rule of law, good governance, transparency, etc. Any threat to these values jeopardises its societal integrity, hence affecting European identity. Similarly, community identity at the national level is built around the feelings resulting from the political affinity amongst its members. This affinity is based on particular features, such as defined geographical space, that can include a set of negotiated aspects, such as the notion of citizenship. The national identity of a group of people who share 'a community of history and culture, possessing a unified territory, economy, mass education system and common legal rights' (Smith 1996) is sometimes regarded as a socially constructed community, which, in other words, is known as a 'political community'. However, within each political community, several cultural communities may exist, defined by ethnicity, for example. These cultural communities are still tied to one broader national culture, forming a single identity as one political community known as 'civic nationalism' (Ipperciel 2007, 397). A cultural community may nevertheless find an affinity towards identical cultures across borders, where shared elements are more strongly connected than that of the ones found within 'civic nationalism'. Understanding that the formation of an identity is not just a deliberate imposition but an inherent development, by extension, a political community or society can also be formed beyond national political structures, within a relatively broad geographical setting. A cultural community can be composed within sub-national as well as cross-national (regional) structures, albeit defined by a set of values at various levels. The basic characteristics of a state – defined territory, government, definite laws and sovereignty – while ensuring a political existence, are not necessarily inclusive enough of the cultural communities within it unless forms of identities of various other groups within it at various levels are acknowledged. A state as a political community receives legitimacy in the exercise of its political function by acknowledging the values inherent in the cultural and communal practices of various other groups. Hence, socially constructed ideas of a nation as a single community is at times problematic.

The Barents Region provides us with examples in this regard. The region is situated in the far North, containing areas that are far from the central capitals of each of the Barents countries. However, the region as a whole is identified as distinct, sharing identical characteristics across borders in terms of environment, demographic composition, livelihood practices and common customary set of norms in the maintenance of everyday life amongst the communities. Thus, an identifiable cultural community in the Barents exists where shared elements are more strongly connected transnationally than that of the ones found within 'civic nationalism'. To the extent that the formation of an identity, as referred to earlier, is not just a deliberate imposition but an inherent development, a new form of political community or society can be formed to extend beyond national political structures, where a group of states recognise a set of common values inherent in themselves. Such a society is defined as follows:

> a group of states, conscious of certain common interests and common values, form a society in the sense that they conceive themselves to be bound by a common set of rules in their relations with one another, and share in the working of common institutions.
>
> (Bull 2002, 13)

The creation of the Barents Euro-Arctic Council (BEAC) – the inter-governmental forum of countries around the Barents Region – can also be considered a political community. Each state acknowledges the common interests and values of the segments of the population that transnationally recognise a common history, common culture, common values, etc. Thus the 'we' feeling, we argue here, is context-dependent. The 'we' feeling in perceiving a national identity may co-exist in the 'we' feeling in perceiving a transnational communal identity. The point is that states' deliberate accommodation of certain ideological features transcending across national borders promotes further legitimisation in its exercise of sovereign function, by which states recognise a transnational society. According to Nelson (2006), although states can theoretically exist through their basic structures – possessing defined territories, governments, definite laws and sovereignty – without an ideology of legitimation, a state loses its authority generally derived from the negotiated understanding between the rulers and the ruled. This sort of political identity building is also connected to 'region building' (e.g., Keskitalo 2004a) via the creation of trans-national institutions. In the case of the Barents Region, the presence of indigenous peoples and their particular context in a regional and societal scope, for example, require the integration of political processes transcending their national border lines.

1.2 Broadening security concept

The traditional understanding of security derives from the Westphalian notion of state sovereignty, driven by political realism, where security discourse is defined through power relations (Elman 2012, 16). Realism calls for military build-up in order to protect the territorial integrity and political independence of nation states using military means as the decisive factor to provide security. There have been serious discussions in the 1980s seeking to include other elements within the security framework such as economic collapse, political oppression, resource scarcity, overpopulation, diseases and natural destruction which also eventually threaten state sovereignty, either from inside or outside of a given territory. However, no clear challenge to state-centric security matured out of such discussions until the Cold War tensions faded.

In the beginning of the 1990s, just at the end of the Cold War, a small group of scholars led by Barry Buzan and his colleagues at the Centre for Peace and Conflict Research in Copenhagen developed a framework to re-conceptualise security discourses by expanding the concept to extend beyond a sole focus on military relations between states (Buzan 1991). Their work established the foundation of securitisation theory by which they broadened the concept to include four other dimensions – political, economic, societal and environmental – in addition to the military dimension. The reconceptualisation of security does not undermine the state centric approach, but is replaced by a broader 'network-based' and 'horizontal' structure, given that the modern threats do not necessarily emanate from states but also from nebulous sources (Burgess and Sissel 2008), such as the actions of non-state actors, or from social phenomena, such as threats arising out of social inequalities. However, the broadening of the concept of security has not been unchallenged. The presence of unrestricted elements identified as possible security threats, such as within the ambit of human security discussions, complicates the security framework (Ullmann 1983, 129). It has nevertheless been argued that, '[c]onceptualizing security lies in the real conditions of insecurity suffered by people and collectives' (Booth 2005, 22) which can be impacted by multiple other sources such as internal conflicts, environmental degradation and economic crisis. The process of horizontal broadening continued to expand to include many other inter-dependent challenges as part of new risk scenarios, hence posing new threats – for example, the emerging cyber security threat.

Hence, the post-realist approach to security is founded on a socially constructed structure whereby security has not only broadened in scope but also deepened to include the subjects facing security threats (Jano 2009, 74; Tarry 1999, 1). To determine the scope of security, Buzan et al. (1998) suggest that a given issue be transformed into a security matter because of its social construction, whereby the transformation is due to social pressure built around that particular referent object identified as an 'existential' threat requiring the adoption of extraordinary or emergency measures. Generally, actors define an issue as an objective threat, which is not automatic but a deliberate process of identification of an object as threat, under the claim that security is a 'self-referential practice: an issue becomes a security issue only by being labelled as one' (Diskaya 2013). Then the referent audience – the subjects of security – is mobilised around the object of threat, and their acceptance of the issue as a serious concern for survival pushes the central actors to introduce emergency measures by adopting changes in policies. A policy change is required to bring the referent object back within the normal process of politics. The acceptance by the audience provides the actors with the legitimacy to use extraordinary means in order to ensure the referent object's removal from the threat agenda (Buzan et al. 1998, 21). As such many more subjects, actors, factors and dimensions at various levels can be included within the security framework, and each of them is, in one way or another, inter-connected and interdependent. While these interdependencies complicate the security picture and present diversified challenges in our everyday lives, all of them do not necessarily begin as or become existential threats. They eventually become existential threats if they are not received as political priorities during their emergence within the normal process of politics.

1.3 Societal security

The theory of securitisation refers to societal security as the protection of 'we' feeling, and requires emergency measures to be materialised for its protection. To the extent that the reconceptualisation of security pertains to survival, a state cannot survive an existential threat to its sovereignty – making it a national security issue within traditional military security. Similarly, when a community is threatened and faces the possibility of losing its 'we feeling' – the identity – it ceases to survive collectively, making the community's existence uncertain in a societal context. A threat to societal security may also result in conflicts between the state authority and the societal community at risk, thereby ultimately affecting state sovereignty and jeopardising state-centric traditional security.

As previously defined, a society is formed through the belongingness of its members in a given community around certain values shared by them as a group, the continued preservation of whose identity/existence is referred to as societal security. It is one of the expanded notions of security, which offers the security of collective groups in relation to other communities or the institutions of the state in which they reside (Thiel 2007). Societal security can also be presented as identity-based self-determination, where a particular community meets its people's needs for belonging to that particular societal formation. According to Buzan, societal security is:

> the ability of a society to persist in its essential character under changing conditions and possible or actual threats. More specifically, it is about the sustainability, within acceptable conditions for evolution, of traditional patterns of language, culture, association, and religious and national identity and custom.
>
> (Buzan 1993, 41–58)

Sustainability within the acceptable conditions for evolution is dependent on many issues in a particular geographical setting and varies by region. The protection of the surrounding environment of a region, for example, provides a community with a platform to exercise its socio-economic and cultural functions, and in which it flourishes in itself as a distinct societal group with particular belongingness. Therefore, when a feeling of belongingness is threatened because society's essential characteristics do not offer the necessary conditions for its continued existence and evolution, the society suffers from the loss of its societal identity. The creation of threats depends on a group's capability of constructing socially powerful arguments that the 'we' feeling is threatened. The construction of collectivities as well as the perception of threats are nonetheless context-dependent. Buzan argues that threats against societal values come from within the states rather than beyond (Buzan 1991, 123). However, at times external forces beyond the national border also play a role whereby they can influence one group of society more rigorously than the other – which makes the distinction between 'within' and 'outside' categories difficult (Ozolina 2016, 15). This is particularly the case in the context of the Barents Region where both transnational forces as well as cross-border issues influence the communities from both inside and outside in various ways. For example, states' national interests sometimes arguably conflict with regional interests and threaten societal security. Yet, the state can also be a promoter of societal security. Unless a balance between these interests is properly addressed, societal security in the regional context becomes an intricate issue. Societal security is threatened when individuals perceive changes as threats to their wellbeing or to community viability. The wellbeing of individuals also depends on the continued functioning of a community as a social group. Such functioning includes factors such as successful resource management, the role of public services and infrastructure (Arsæther 2004), the possibility of participation in policy making (Duhaime and Caron 2009), and a set of common values from which most individuals derive their identity (UNDP 1994).

In addition to a broadening theoretical analysis of societal security, the practical approach to societal security has adopted an interdependent and broader, conceptual framework. Today, the broadened scope of security is linked to issues found also in safety discourses, such as in emergency preparedness and response mechanisms. The inter-relationship between security and safety is a relatively recent development in relation to societal security. Both safety and security are concepts that are intertwined in understandings of human challenges. The aim is to utilise a bridge between security and safety concepts to deal with challenges that have consequences for societies both within, and transcending, the national border. Bengt Sundelius formulated the security/safety inter-linkages to refer to societal security (Sundelius 2005). Moreover, the complexity of modern society requires an understanding of our increasing interdependencies with new technologies. Today, infrastructural interconnectedness has become an obvious part of our daily lives, and society is dependent on the reliable functioning of the system found within this interconnectedness (Sundelius 2016, 160). For example, uninterrupted functioning of IT networks is connected to uninterrupted energy supply. A breakdown of one of the systems affects the other one and vice versa. Therefore, the theoretical approach to societal security presented in the Copenhagen School as social construction is somewhat different from what it refers to at the practical level.

A definition of societal security is presented in the International Organization for Standardization (ISO/TC 223). According to the definition, societal security refers to 'protection of society from and response to incidents, emergencies, and disasters caused by intentional and unintentional human acts, natural hazards, and technical failures' (Tangen 2011). The definition also highlighted threats in connection to all-hazards perspective 'covering adaptive, proactive and reactive strategies in all phases before, during, and after a disruptive incident'. According to the definition, societal security 'is multidisciplinary, and involves actors from both the public and private sectors, including not-for-profit organizations'. Given that challenges threaten communities differently, an approach to ensuring societal security needs to be rather flexible both in scope and in scale to take into account the complex, yet often intertwined, challenges that various communities and community members experience (see also Hossain et al. 2017). The diversity of threats to societal security is linked to the framework of human security – a policy tool to analyse diverse perspectives of security threats at the individual and community levels. Hence the notion of human security, as discussed in the following section, can be used as an effective analytical tool to interpret societal sustainability within acceptable conditions.

1.4 Linking human security to societal security

The core notion of human security is not new. Its origin can be traced back to the 17th century social contract theory, where individuals agree around a number of common security goals to be provided by an entity they approve themselves – the so-called sovereign entity. However, the modern approach to human security, re-defined since the early 1990s, rejects hegemonic relationships between states and its subjects and calls for a broader inclusive agenda. The Human Development Report (HDR) of the United Nations Development Program (UNDP) (UNDP 1994) endorsed the concept of human security in order to safeguard the humans (and communities) at the sub-state level. The concept has been used to refer to situations or conditions which affect the survival of individuals, and not that of the states'. The referent objects of security include individuals; it also includes communities such as ethnic minorities, and indigenous and local communities, who are largely the victims of human security challenges in diverse ways. The security threats to these referent objects, unless safeguarded, may, however, result in large-scale societal unrest that has the potential to disrupt traditional national security. Therefore, at times the concept of human security can be understood as an expansion of traditional security concerns that put individuals at their core. Going beyond physical security, it perceives a holistic relationship to socio-economic, environmental and cultural wellbeing, where an individual can freely maintain and develop their identity.

The HDR incorporated seven categories of threats to human security: environmental, economic, food, health, political, personal and community security. These categories relate to concerns associated with violent conflicts from various sources at both the intra-state and supra-state levels. They also largely relate to non-violent issues affecting the conditions of human lives, such as diseases, hunger, environmental catastrophes, etc. Such non-violent aspects of security have been reported to 'kill far more people than war, genocide, and terrorism combined' (Human Security Centre 2005). Protection from threats and the promotion of identified objects of the security agenda are the goals of human security and seek to broadly ensure 'freedom from fear' and 'freedom from want' among the individuals and communities in a given society. While the former is about guaranteeing security from physical violence, the latter concerns economic, social and cultural demands of individuals and communities (UNDP 1994). Human dignity suffers where such fulfilment is not guaranteed. Since the causes and expressions of human security depend on a complex interaction of regional and local factors, the concept offers normative and procedural tools to respond to particular threats arising from the interactions affecting overall societal structure in a given geographical setting. As a broad, dynamic and flexible framework, human security captures such variation in specific contexts; builds processes that are based on people's own perceptions of fear and vulnerability; identifies the concrete needs for people under stress; enables solutions based on local realities; helps identify priority security needs at the local level; and looks at the impact of global developments on different communities (HSU 2009). By providing a holistic and contextual account of the concrete needs of individuals and communities and the factors endangering their security, human security, when applied in specific context, is capable of analysing and assessing the overall social impact (ibid.). The concept of security is negotiated and formed amongst the various stakeholders at various levels. Therefore, by integrating a larger agenda, such as economic globalisation, the environment and sustainability into the concept of security, the identification of threats and the mechanisms for their removal, by extension, promotes the self-preservation of societal communities.

Human insecurities are closely related to environment-induced phenomena that influence many factors affecting the lives of individuals and communities. Such diverse insecurities can hamper societal cohesion, which then threatens to destabilise society's political structure. Political instability leads to various societal challenges, such as resource scarcity and social discrimination (Dabelko et al. 2000). We, by highlighting the concept of human security, identify societal challenges and thereby we go beyond the securitisation approach of existential threats. We argue that societal security is not just a fight for survival, it is also about (human security's approach to) the promotion of conditions for the greater sustenance of a society – it is about 'survival plus' (Booth 2007, 102). It is about any danger of risks or threats that may have the potential to affect the referent object's survival but that may not be mature enough to threaten society's immediate existence. While survival refers to an existential condition, societal security perceives an ability to promote socio-cultural, economic and environmental abundance for society's continued sustainability. Often, this aspect of security is referred to as 'positive security' – in other words, 'security to' (enablement) rather than 'security from' (threats) – found in human security literature (Hoogensen 2012, 836). Addressing security this way accommodates the everyday concerns of a given community, making it more relevant, recognised, visible and socially integrative (ibid., 838), which eventually promotes an ability to make 'life-choices' for that particular community (Booth 2007, 106). Thus, we highlight the human security concept's normative utility and guiding role for the promotion of value additions and ethical norms in security debates (Tadjbakhsh and Chenoy 2007) as they relate to the formation and promotion of societal sustainability.

1.5 Societal security: the Barents Region in context

The Barents Region transcends the national borders of four countries. The region comprises the Arctic and sub-Arctic regions of mainland Europe, encompassing the northern parts of Norway, Sweden and Finland, and north-western Russia. The region is the homeland of around 5.2 million inhabitants, including a number of indigenous groups. The majority of the Barents population live in the Russian parts of the region. While the people of the region are (sub-)nationally divided, most people live in relatively large urban areas. However, the population size in these places is small (Tennberg et al. 2014). They are composed of mixed groups of local and indigenous peoples, forming smaller communities within each group and speaking different languages. These smaller communities are also tied to other groups and to the region through shared experiences. These include the enjoyment of unique cultural and communal distinctness derived from common interests developed in connection to common geo-physical structures supported by identical natural environments, similar cultural practices and values, and common threats affecting their lives, livelihoods and community wellbeing. They find likeness amongst themselves beyond national dimensions in many ways. For example, the Sámi live in four countries in the region transnationally, yet share a common understanding as it relates to their cosmology and unique worldview, which is distinct from those of other groups of people living outside of the Barents Region in these countries. This likeness extends also to other groups of peoples, not just indigenous peoples in the Barents, and has led to the formation of a broader identical community known as the Barents community, which has been recognised and actively promoted through various political processes. The creations of the Barents Euro-Arctic Council (BEAC) in 1993 and the Barents Regional Council (BRC) within the framework of BEAC are examples of transnational region building where the common concerns led to a formalisation of the region in institutional structures (Keskitalo 2004b). The process started at the end of the Cold War in 1991 with the establishment of the Protection of Arctic Environmental Strategies (AEPS) (later integrated within the framework of the Arctic Council), where the integration of common environmental concerns were incorporated in a broader Arctic framework (e.g., Koivurova 2011). These concerns, as they affect the whole of the Arctic region, and as they develop over time, have been identified as risking society's unique features in many ways. The Arctic Human Development Report (AHDR) has largely addressed many of these issues from the viewpoint of socio-cultural, economic and environmental aspects (Larsen and Fondahl 2014) in order to promote community cohesion within the region.

Many of these challenges, as they affect societal security in one way or another, are discussed in detail from the perspective of human security in the subsequent chapters of this book. However, in order to better understand the particular issues discussed in those chapters, we herein provide a general overview of the conditions prevailing in the Barents, which influence the threats and opportunities present in this transnational society.

The Barents cannot be identified as a completely homogenous region; hence its population cannot be identified as one homogenous community in any strict sense. As discussed, it includes many small groups of communities in each of its sub-regions that share similarities and differences. Political and social variations in each of these sub-regions (which are split into a number of territories within different countries) offer diverse experiences for their inhabitants. For example, the sub-regions of the Russian Barents cannot be comparable to those of the Nordic Barents Region. These sub-regions are different from each other across countries, even sometimes within one single country – some places are more infrastructurally developed and some are less; some are located in remote and inhospitable places and some others are more easily accessible; and the population in some places are subject to more political discrimination than others because of varied political infrastructure, etc. As shown in the chapter on economic security, human development indicators vary widely in different parts of the Barents Region. Despite this variance, determining the formation of society does not depend entirely on narrowly defined indicators. Common challenges, as indicated, have the potential to influence community structures and perceived values, albeit at varied levels and varied scales, and to create an identical society. The uniqueness of this societal formation is that it extends beyond civic nationalism and towards identical elements shared amongst members, regardless of their nationality.

Thus, the uniqueness of the Barents is that the region's geographic setting, as well as climatic conditions, are largely identical. Its unique natural environment provides the population with a similar means of livelihood and subsistence, offering a form of transnational homogeneity in a broader sense. Today, the transformation of the regional environment presents common challenges across the Barents community in terms of livelihood practices, modifications to regional economies, alterations in demographic balances, changes in socio-cultural values, etc.

For example, the effects of climate change are rapidly impacting the region, and the resulting increase in temperature and the expectation of future warming presents a common challenge for Barents states to address (Kattsov and Källén 2005; Overland et al. 2011). While climate change and its consequences pose both new challenges and new opportunities in the Barents Region, the region's environmental and societal sustainability is often times imbalanced due to uncertain and unsafe societal conditions (Mustonen and Nieminen 2007) that commonly affect the inhabitants of the region. The region also similarly faces the consequences of increasing economic globalisation, as numerous economic activities, such as forestry, oil and gas, mining and related extractive industrial activities, energy development, tourism, etc. (Tennberg et al. 2014), replace the traditional economic activities, such as hunting, fishing, reindeer herding, berry and mushroom picking, etc. While the former is of importance for national, local and regional economies (Glomsrød and Aslaksen 2009), the latter is relevant for environmental sustainability, local economy, subsistence and recreation of communities in the region (Jansson et al. 2015, 32).

Whereas there is a need for the co-existence of both economic and environmental sustainability for the region's stability, research shows that the benefits derived from economic development are not shared equally, giving rise to social inequality and leading to social injustice and societal insecurity. Conflicts of different interest groups over land use and over the unequal distribution of resources often pose threats to socio-economic opportunities. The expansion of extractive industrial activities and their subsequent impact jeopardises the socioeconomic and environmental integrity of the region. Local inhabitants suffer from increasingly uncertain living conditions whereby they lose possibilities for economic subsistence as a group. Moreover, the cultural sustainability of the region is also at risk due to patterns that increasingly lead to demographic changes. Tennberg states that one common characteristic across the Barents Region is out-migration (Tennberg et al. 2014), which impacts the region's societal structure in terms of demography. While out-migration of the young population is observed in many parts of the region, an inflow and increase in the population are also evident in many other parts. For example, as the opportunity to practise traditional activities decreases and opportunities grow from increased extractive activities, people from outside the region are attracted to move to the region. While this is more or less common across the region, it is most common in the Russian Barents Region (Heleniak and Bogoyalensky 2014). A transformed societal structure, with newcomers in the region possessing different skills and education (Rasmussen et al. 2014) and with the usage of new technology reducing the differences between local and global practices in everyday life, influences the culture and tradition of existing communities. Hence a transformed societal structure offers a new lifestyle having had influence, among others, on the local languages spoken amongst the communities, which is, in fact, vital for the maintenance of cultural identity. For example, in the region, elderly people often speak indigenous languages mostly, while younger generations often grow up bilingual, with the national language as the dominant one (Schweitzer et al. 2010).

The consequences of such changes present a broad threat and put the region's indigenous populations at risk, which eventually leads to denial of a coherent societal progression. Therefore, necessary and timely response mechanisms should be developed for the continued viability of societal functions for greater stability. Our perception of societal security in the context of the Barents Region, as referred to earlier, addresses a broader conceptualisation of security that extends beyond the limited scope endorsed within traditional securitisation theory. These new risk landscapes, as referred to by Burgess and Sissel (2008), are constitutive of societal security threats, and we have taken them into consideration. For Burgess and Sissel, societal security 'is not connected to specific threats or scenarios but encompasses all possible strains and hazards that a society could face' (Burgess and Sissel 2008). Thus, societal security is connected to 'objective factual threat', which forms security in relation to 'disaster risk reduction', 'resilience of nations and communities to disasters', 'civil protection' and 'civil emergency planning'. These aspects of security are different in nature from a socially constructed security approach, such as those articulated in the Copenhagen School's conceptualisation of securitisation. The promotion of security refers to developments of new and rapid response mechanism in reply to these new risks (ibid.). Thus, we find that the concept of human security, as it applies to the region, offers effective mechanisms for both the protection and promotion of societal security.

The response mechanisms associated with human security offer measures that are applicable in both reactive and proactive manners in order to promote society's sustainability within 'acceptable conditions'. Given the indeterminate nature of the term 'acceptable conditions' as it relates to societal sustainability, the process-oriented approach to security, found within the analytical framework of human security, pertinently fits to create guiding norms for societal security discussions. Therefore, protecting or maintaining society's essential features within acceptable conditions for evolution requires a process, which is best analysed by human security tools. Such a process has been referred to by Ozolina (2016, 10) as 'securitability'. The concept of 'securitability', as it applies to societal security, not only relates to protection from threats but also perceives both promotion of resilience and empowerment of individuals and communities going hand-in-hand with the concept of positive security as discussed elsewhere in this chapter. Resilience refers to the promotion of 'the capacity of the system to continually change, adapt, and yet remain within critical thresholds' and 'the ability of an individual to recover from adversity' (Berkes and Ross 2013, 6). Empowerment, on the other hand, is about the promotion of the involvement of individuals and societal networks to proactively invoke their personal and communal choices (Ozolina 2016, 10). Securitability provides a process that 'attempts to advance political priorities in the face of a new threat landscape' (Sundelius 2016, 162) resulting from, for example, threats to food, health, pollution and climate change. The functionality of a society is influenced by such attempts, which eventually promote broader human wellbeing that contributes to communal viability. The term 'wellbeing', like 'acceptable conditions', cannot be precisely measured by one metric. Thus, a human security approach provides guidance for measuring wellbeing by taking specific indicators and perspectives into consideration. The specific indicators offer an impetus on how each of them is relevant to analyse threat pictures in a given part of the region.

Part 2 of this book has extensively analysed a number of identified human security factors as they relate to the Barents Region and provided recommendations to guide the use of human security in promoting societal security. In particular, methods prescribed by the concept of human security – the identification of specific challenges in a given geographical context, the adoption of strategies indicating effective measures that integrate the concerns and voices of relevant actors, and the implementation of these strategies by invoking established and emerging norms such as resilience – can be endorsed as the guiding norm for the promotion of societal sustainability. The Barents Region faces multiple stresses, both internal and external, natural and human-induced, and which have raised social, political and environmental concerns (Adger 2000, 347–64). These concerns relate to the economy, social capital, formal and informal relations amongst the communities, the promotion of skills via training and education, as well as the creation of knowledge for information and communication in emergencies (Jermalavicius 2015, 160–61). To this degree, they impact the ability of a society to maintain sustainability within acceptable conditions. Such acceptable conditions within the Barents Region require an ability to manage outside influences and local culture and social competence (Rasmussen et al. 2014, 462). Our arguments in this book lie in the recognition of societal security as central to resilient security strategies. Societal security underpins the development of abilities to respond to threats to communities, regardless of whether such threats are yet to mature, emerging or existential.

References

Adger, W.N. 2000. Social and ecological resilience: are they related? *Progress in Human Geography* 24(3): 347–364.

Arsæther, N. 2004. *Innovations in the Nordic Periphery*. Stockholm: Nordregio.

Berkes, F. and Ross, H. 2013. Community resilience: toward an integrated approach. *Society & Natural Resources* 26(1): 5–20.

Booth, K. 2005. Introduction to Part One. In *Critical Security Studies and World Politics*, edited by K. Booth. London: Lynne Rienner.

Booth, K. 2007. *Theory of World Security*. Cambridge: Cambridge University Press.

Bull, H. 2002. *The Anarchical Society: A Study of Order in World Politics*. New York: Columbia University Press.

Burgess, J.P. and Sissel, H.J. 2008. The influence of globalization on societal security: the international setting. *PRIO Policy Brief* 3.

Buzan, B. 1991. *People, States & Fear: An Agenda for International Security Studies in the Post-Cold War Era*. London: Harvester Wheatsheaf.

Buzan, B. 1993. Societal security, state security and internationalization. In *Identity, Migration and the New Security Agenda in Europe*, edited by O. Wæver. London: Pinter. Buzan, B., Wæver, O. and de Wilde, J. 1998. *Security: A New Framework for Analysis.* London: Lynne Rienner Publishers.

Cambridge Dictionary. n.d. http://dictionary.cambridge.org

Dabelko, G., Lonergan, S. and Matthew, R. 2000. *State-of-the-Art Review on Environment, Security and Development Cooperation.* Paris: OECD Development Assistance Committee.

Diskaya, A. 2013. Towards a critical securitization theory: the Copenhagen and Aberystwyth Schools of Security Studies.' *E-International Relations*, www.e-ir.info/2013/02/01/towards-a-critical-securitization-theory-the-copenhagen-and-aberystwyth-schools-of-security-studies/

Duhaime, G. and Caron, A. 2009. Economic and social conditions of Arctic regions. In *The Economy of the North 2008. Report 112*, edited by S. Glomsrød and I. Aslaksen. Oslo: Statistics Norway.

Elman, C. 2012. Realisms. In *Security Studies: An Introduction*, edited by P.D. Williams. London: Routledge.

Glomsrød, S. and Aslaksen, L. 2009. *The Economy of the North 2008*. Oslo: Statistics Norway.

Heleniak, T. and Bogoyalensky, D. 2014. Arctic populations and migration. In *Arctic Human Development Report II. Regional Processes and Global Linkages*, edited by J. N. Larsen and G. Fondahl. Copenhagen: Norden.

Hoogensen, G. 2012. Security by any other name: Negative security, positive security, and a multi-actor security approach. *Review of International Studies* 38(4): 835–859.

Hossain, K., Zojer, G., Greaves, W., RonceroJ.M. and Sheehan, M. 2017. 'Constructing Arctic security: an inter-disciplinary approach to understanding security in the Barents Region'. *Polar Record* 53(1): 52–66.

HSU. 2009. *Human Security in Theory and Practice. An Overview of the Human Security Concept and the United Nations Trust Fund for Human Security.* Human Security Unit, United Nations. www.un.org/humansecurity/sites/www.un.org.humansecurity/files/human_security_in_theory_and_practice_english.pdf.

Human Security Centre. 2005. *Human Security Report 2005: War and Peace in the 21st Century.* New York: Oxford University Press.

Ipperciel, D. 2007. Constitutional democracy and civic nationalism. *Nations and Nationalism* 13(3): 395–416.

Jano, D., 2009. Aspects of security 'dilemma': what we have learned from the macedonian case. *Perceptions Journal of International Affairs* 14: 73–90.

Jansson, R., Nilsson, C., Keskitalo, E.C., Vlasova, T., Sutinen, M.-L., Moen, J., Iii, C., Stuart, F., Bråthen, K.A., Cabeza, M., Callaghan, T., van Oort, B., Dannevig, H., Bay-Larsen, I., Ims, R. and Aspholm, P.E. 2015. Future changes in the supply of goods and services from natural ecosystems: prospects for the European north. *Ecology and Society* 20(3): 32.

Jermalavicius, T. 2015. Societal resilience as a deterrent in 'hybrid war'. In *Riga Conference Papers 2015: Towards Reassurance and Solidarity in the Euro-Atlantic Community.* Riga: Latvian Institute of International Affairs.

Kattsov, K. and Källén, E. 2005. Future climate change: modelling and scenarios for the Arctic. In *Arctic Climate Impact Assessment Scientific Report*. Cambridge: Cambridge University Press.

Keillor, B.D., Hult, G.T.M., Erffmeyer, R.C. and Babakus, E. 1996. NATID: the development and application of a national identity measure for use in international marketing. *Journal of International Marketing* 42(2): 57–73.

Keskitalo, E.C.H. 2004a. *Negotiating the Arctic: The Construction of an International Region*. New York and London: Routledge.

Keskitalo, E.C.H. 2004b. The Arctic as an international region – but for whom? In *Arctic Governance*, edited by T. Koivurova, T. Joona and R. Shnoro. R. Juridica Lapponica 29. Rovaniemi: Oy Sevenprint.

Koivurova, T. 2011. Environmental protection in the Arctic and Antarctica. In *Polar Law Textbook*, edited by N. Loukacheva. Copenhagen: Norden.

Larsen, J.N. and Fondahl, G. 2014. *Arctic Human Development Report Regional Processes and Global Linkages*. Copenhagen: Nordic Council of Ministers.

Mustonen, T. and Nieminen, M. 2007. Knowledge of snow, weather and the landscape – snowchange years with the Sámi. *Barents watch*. www.bioforsk.no/ikb Viewer/Content/96985/BW07_engelsk_nett%20(2).pdf.

Nelson, B. 2006. *The Making of the Modern State – A Theoretical Evolution*. New York: Palgrave Macmillan.

Overland, J.E., Wood, K.R. and Wang, M. 2011. Warm Arctic – cold continents: climate impacts of the newly open Arctic Sea. *Polar Research* 30(1): 1–14.

Ozolina, Ž. 2016. Societal security: conceptual framework. In *Societal Security Inclusion-Exclusion Dilemma. A Portrait on Russian Speaking Community in Latvia*, edited by Ž. Ozolina. Riga: Zinatne Publishers.

Rasmussen, S.O., Bigler, M., Blockley, S.P., Blunier, T., Buchardt, S.L., Clausen, H.B., Cvijanovic, I., Dahl-Jensen, D.*et al.*2014. A stratigraphic framework for abrupt climatic changes during the Last Glacial period based on three synchronized Greenland ice-core records: refining and extending the INTIMATE event stratigraphy. *Quaternary Science Reviews* 106: 14–28.

Ross, M.H. 2009. Culture in comparative political analysis. In *Comparative Politics: Rationality, Culture, and Structure*. New York: Cambridge University Press.

Schweitzer, P., Irlbacher Fox, S., Csonka, Y. and Kaplan, L. 2010. Cultural well-being and cultural vitality. In *Arctic Social Indicators: A Follow-up to the AHDR*, edited by J.N. Larsen, P. Schweitzer and G. Fondahl. Copenhagen: Norden.

Smith, A. 1996. The origins of nations. In *Becoming National: A Reader*, edited by G. Eley and R.G. Suny. New York: Oxford University Press.

Sundelius, B. 2016. Societal security: an emerging field of scholarship underpinning practices in the Baltic Sea Region. In *The Baltic Sea Region: Hard and Soft Security Reconsidered*. Riga: Latvian Institute of International Affairs.

Sundelius, B. 2005. A brief of embedded societal security, information & security. *An International Journal* 17: 23–37.

Tadjbakhsh, S. and Chenoy, A.M. 2007. *Human Security: Concepts and Implications*. Oxon: Routledge.

Tangen, S. 2011. *Revised Business plan for ISO/TC 223, Societal security*. ISO/TC 223 Societal Security.

Tarry, S. 1999. 'Deepening' and 'widening': an analysis of security definitions in the 1990s. *Journal of Military and Strategic Studies* 2(1): 1–13.

Tennberg, M., Vola, J., Espiritu, A.A., Fors, B.S., Ejdemo, T., Riabova, L., Korchak, E., Tonkova, E. and Nosova, T. 2014. Neoliberal governance, sustainable development and local communities in the Barents Region. *Barents Studies* 1: 41–72.

Thiel, M. 2007. Identity, societal security and regional integration in Europe. *Jean Monnet/Robert Schuman Paper Series* 7(6): 1–15.
Ullmann, R.H. 1983. Redefining security. *International Security* 8(1): 129–153.
UNDP. 1994. *Human Development Report 1994*. New York, Oxford: Oxford University Press.

1.2 The Barents Region, a society with shared security concerns in the Arctic

Dorothée Cambou and Lassi Heninen

Introduction

Since the beginning of the 19th century, the Barents Sea area, as a sub-region of the entire Arctic, has been considered one of the most peaceful regions in the world. However, historically, there was high tension between the countries of the northernmost parts of Europe – Finland, Norway, Sweden and Russia – and the relationships between these counties have been characterised by their struggle for sovereignty and security.[1]

Despite historical tensions and conflicts, in 1993, Finland, Norway, Russia and Sweden established the so-called Barents Euro-Arctic Region – covering the northernmost counties of Finland, Norway and Sweden, and the North-western regions of the Russian Federation – based on common interests, and the desire to strengthen shared identities across national borders (e.g., Heininen 2009). With the Barents cooperation in place, interregional cooperation between the northern parts of the Nordic countries and Russia drastically improved over the years and allowed for the creation of a governance framework to improve the human security of its inhabitants. As mentioned by Former Norwegian Minister of Foreign Affairs, Bjorn Thore Godal,

> throughout most of history … a regional identity emerged, a sense of common destiny among peoples sharing the experience of trying to make a living in harsh surroundings.
>
> (Godal 1996)

In practice, this has contributed to foster peace and stability in the region and laid the ground for the creation of a political region. Today, despite growing tension between Russia and the West, specifically the USA and European Union, and the consequent decrease in activities of Arctic cooperation since 2014 as a consequence of the crisis in Ukraine, cooperation in the Barents Region still stands strong in promoting the wellbeing of the inhabitants of the region.

Against this background, it is the aim of this chapter to describe and analyse inter-regional cooperation in the European Arctic. More specifically, the objective of this chapter is to provide a contextual background describing the major characteristics that led to the establishment of the Barents cooperation. While the assessment of this development for improving human security in the region is mostly left to the following chapters, the last section concludes with some remarks about the specific identity of the Barents Region as a special 'community' within the Arctic Region, as well as the importance of the interregional cooperation to strengthen human security in the region.

1 Geography and demography of the Barents Region

The Barents Region covers an area of 1.75 million square kilometres (about three times the size of France), which extends from Lofoten in Norway in the west to the Russian coal-mining town of Vorkuta in the east, and from Lake Ladoga in the Russian Karelia in the south to Nordkapp in the North.[2] The Barents Region includes lands and territories situated in the northernmost parts of the continental Europe on the southern coast of the Barents Sea, and consists of the northernmost parts of Norway, Sweden and Finland, and the northwestern regions of the Russian Federation. The Barents Sea, named after a Dutch explorer, Willem Barentsz who undertook three Arctic expeditions searching for the North West Passage in the sixteenth century, does not, however, include the official Barents Region. As a political region, the Barents

Figure 1.2.1 Maps of the Barents Region by the Barents Norwegian Secretariat

area comprises the land along the coast of the Barents Sea but does not include cooperation over the sea. This is mostly due to unsettled sovereignty issues between Norway and Russia in the Barents Sea (Carrillo 1998, 21).

From a geographic perspective, the Barents Region constitutes parts of the Arctic and sub-Arctic regions which share a number of physical features. The Kirkenes Declaration (1993) which created the Barents Euro-Arctic Council notes that these include a 'harsh climate, sparse population and vast territory'.[3] To this could be added physical features such as large swathes of Arctic tundra, polar night and midnight sun and areas of permafrost. There are however, a number of differences, particularly between the Nordic countries and Russia, such as living standards, language, culture, religion, history and political traditions (Zimmerbauer 2012, 94). The Barents Region is located at the intersection between eastern and western culture, between Catholic and Orthodox Christianity (nowadays Lutheran and Orthodox), and consequently, the population also shares different languages, cultural and religious traditions. Furthermore, it is a region that is culturally diverse with different livelihood systems.

The Barents Region is a home to 5.23 million people, the Finns, Norwegians, Russians, Swedes and a number of indigenous people groups. Among the inhabitants of the region, the Sami, the Nenets and the Veps account for the indigenous peoples living in the region.[4] These peoples have much maintained their traditional way of life with different usage of local resources, which were established before nation-states were formed. The most populous of them, the Sami people, are comprised of approximately 70 000–90 000 people living across Finland, Norway, Sweden and the Kola Peninsula, Russia. The Sami communities remain closely attached to their traditional livelihoods including fishing, hunting, reindeer herding and berry picking. It is estimated that 10 per cent of the Sami population is still involved in herding-grazing activities on a fulltime basis. Indeed, their cultural, economic and political development remains a core issue in the region, and for this reason, the protection of the lands and resources of the traditional Sami territory – Sapmi – across the northern borders of the four states of Finland, Norway, Sweden and Russia are of utmost importance. In all three Nordic countries, the Sami people have their own assemblies to represent their interest at the state level and in Russia they are recognised as indigenous peoples. Despite state borders, the Sami people have maintained a common history, culture, language and traditions, and their relations have increased since the end of the Cold War through the development of political and cultural cooperation.

In addition, the Veps, who live in the southern part of the Republic of Karelia and in remote parts of the Leningrad and Vologda Oblasts, form another indigenous minority of approximately 8200 people according to the 2002 census (barentsinfo.org, 2017). While agriculture was always at the heart of their livelihoods, most of the traditional occupations of the Veps have today vanished (Kolga 2001). Finally, the Nenets are the most numerous indigenous people in the Russian Federation. In the Barents Region they form a minority of about

41 000 people living in the Nenets Autonomous Okrug, Arkhangelsk Oblast, Komi Republic and Murmansk Oblast. Today, the majority of the Nenets live in rural communities and are engaged in agricultural sector, in education and health care. In addition, 14 per cent of the Nenets work in reindeer herding, which are still migrating while other have settled in villages. On the whole, the rural population of the Nenets has to count on local renewable resources to fulfill their needs, though it is said that 'the local economic situation has improved thanks to high revenues from oil and gas development' (Tuisku 2004).

In this context, national integration accompanied by modernisation and industrialisation processes have raised tensions and conflicts of interests between peoples and their livelihoods at regional and local levels. While parts of indigenous peoples groups often follow a traditional lifestyle, engaging in activities such as reindeer husbandry and subsistence fishing, there is also a highly skilled workforce in the region, which exploits the rich forestry, mineral, oil, gas, fishing and even diamond resources.[5] It is significant that not many other parts of Europe have access to the amount of forests, fish, oil and gas and other minerals present in the region. This has raised both growing concerns regarding the environment, and has sparked important economic and business interests in the region. Large mining sites have been exploited for decades in the Barents Region. Recently, the region has also received interest due to the discovery of several important natural gas and oil deposits in the Barents Sea and Petsoran Sea.

All in all, the Barents Region is extremely rich in minerals and while extractive industries cause controversy through their impact on the environment and livelihoods of local communities, they are important for the development of the region. Finally, renewable resources are also important, as Nordic counties of the Barents Region have large hydropower resources and are currently investing in massive wind power energy projects to meet the demand for sustainable energy production and consumption.[6]

All these elements constitute the foundations for economic, social and cultural development in the Barents Region and the markers for its specific identity.

2 The history of the Barents Region[7]

The history of the European Arctic, today the Barents Region, is that of colonisation, state expansion and cooperation. Until the 13th century the states' influence was small, and the region 'was uncharted territory totally without borders, taxation, conscription and other traits of nation states' (Gyllenhaal 2017). Between the 13th and 20th centuries, the region was subject to many geopolitical changes, mostly caused by hegemonic power struggles between different sovereigns. Only indigenous peoples, mostly nomadic, and small groups of hunters and fishermen from the Nordic countries and Novgorod lived within the region. Most of these groups adapted to the environmental conditions, possessed different cultures and interacted to some extent with each other. The absence of any homogenous population, still today, raises some difficulty for addressing the history of the Barents Region. Yet, the impact of the colonisation process, the conflicts between nation-states and the establishment of cooperation across state borders have forged a unique historical relationship for the inhabitants of the region, which have helped to shape the contours of a distinct society.

The colonisation of the Arctic region approximately began when non-indigenous populations moved in the region and accentuated trade and cultural relations with local communities. During the 14th century, the local population also became the target of royal taxation and the influence of the church increased. The colonisation process led to the establishment of competing areas of authorities between local settlements and the kingdoms of Sweden, Norway and Novgorod.[8] For indigenous communities, the colonisation of the Arctic also meant their connection to national political structures, including rules of taxation, laws and control over local resources. The colonisation process triggered important changes in the governance of the region, accompanied by major societal transformations fuelled by trade, industrial and cultural developments. Historically, this process accelerated in the more accessible Barents Region during the 16th century while, by comparison, the Canadian Arctic remained relatively isolated until the mid-1900s.

With a focus on territorial control as a key element to strengthen sovereignty, the expansion of nation-states also spurred conflict in the region. Between the 13th and 20th centuries, the countries that today consist of Norway, Sweden, Finland and Russia were involved in various struggles for supremacy in the governance of the population, land and resources of the region. From 1397 until 1523 all the Nordic countries and nations were governed under one kingdom, the Kalmar Union. During part of this period, Sweden was a European great power. From 1523–1814 Nordic governance consisted of two main states. Denmark ruled Norway, Iceland, Greenland and the Faroe Islands, and Sweden ruled Finland from the 12th century until Russia finally won control in 1809. In 1814, Norway became part of a union with Sweden, lasting until 1905 when Norway gained independence. In 1917, the October Revolution created Soviet Russia. In the same year, Finland was able to use the chaos caused by the First World War and the Bolshevik revolution to declare its full independence. The Second World War, with human catastrophes and losses of inhabitants, meant extensive geopolitical changes in the Barents Region. Despite two wars, Finland remained independent but was partly destroyed and lost large territories to Russia, such as the Petsamo region, which was only regained in 1920. Norway managed to keep its independence as well, even though Nazi Germany occupied the whole country and used it, as well as Petsamo, as a convenient launching point for the failed attack on the Russian city of Murmansk. The Soviet Union was the main target of Nazi Germany in the European Arctic. Supported by the Allies and their material assistance, such as maritime escorts to Murmansk, the Soviet Union survived the siege, though with important casualties.

During the Second World War, the northernmost areas of Europe became a place of great geopolitical significance, especially as the German troops occupied Norway and Svalbard, and were in charge of the Northern Front of Finland (comprising half of the Finnish territory). With the Finnish-Russian peace treaty ceding the Petsamo border region to Russia, Norway found itself beside a closed border with the Soviet Union and with a large military presence next door. After the Second World War, the Kola Peninsula was industrialised and militarised, transformed into a platform for industrial and military activities with a stockpile of both conventional and nuclear weapons as an important part of the military competition between the USSR and the USA. As a consequence, traditional interaction in the northern areas became even more difficult, especially considering that there was almost no access across the border from Finland or Norway into or out of the Soviet Union from the late 1940s until the fall of the Soviet Union. This situation consequently underlined an East–West divide that was practically impossible to overcome until the end of the Cold War.

At the same time, Nordic states began to increase trans-border cooperation to strengthen the integration and development of their northern areas. In 1952, the Nordic Council was established. This inter-parliamentary institution has been instrumental in facilitating contact between peoples and developing a sense of affinity between them.[9] The Nordic Council's first real contribution was the establishment of a common labour market and a free visa zone for citizens. In order to further cooperation between Nordic states, the intergovernmental forum of the Nordic Council of Ministers was also established in 1971 to complement the activities of the Nordic Council. In 1967, the northernmost counties of Finland, Norway and Sweden also established the North Calotte Committee as a forum for Nordic cooperation in the North Calotte, to further cross boundary exchanges and cooperation in the field of regional policies, local economic development, culture and art.

Following stronger cooperation between Norway, Finland and Sweden, their collective relations with the Soviet Union also evolved. Although tensions between the two blocs remained, possible modes to integrate the Soviet border regions within the inter-regional cooperation of the North Calotte region were discussed. Starting in 1964, cultural meetings such as the North Calotte Peace Days were organised under the auspices of peace and friendship, and to create a nuclear-weapon-free zone in the region. Various forms of cultural co-operation and the increase of individual contacts also accompanied such meetings across the Iron Curtain (Elenius et al. 2015, 341–343; also Heininen 1999a, 1999b, 107–198).

3 The forerunners of the Barents cooperation: peace and environmental collaboration

In 1985, perestroika opened a new window of opportunity to facilitate cooperation across the East–West border. In October 1987, the Soviet leader

Mikhail Gorbachev gave a speech in Murmansk and proclaimed the Arctic a zone of peace. This event was fundamental to the transformation of the region, constituting the beginning of a new phase in the history of its population. As a result of these new opportunities, the organisation of international peace and environmental events, many of which were Nordic-Russian, took place in the summer of 1988 in Murmansk. Thus, in spite of the Cold War cleavage, and due to the rich tradition of regional cooperation such as the Pomor Trade, 'it was during this period that the seeds of the future Barents were sown' (Elenius et al. 2015, 365) and that cooperation in this sub-arctic region begun its journey.

In his speech, Mikhail Gorbachev (Pravda 1987) made concrete initiatives outlining a goal to establish a nuclear-free zone and 'zone of peace' in the (European) Arctic. Among the initiatives were measures that included a restriction of naval and air force activities in the Baltic, Northern, Norwegian and Greenland Seas and the promotion of confidence-building measures in those areas, cooperation on resource development, the organisation of an international conference on Arctic scientific research coordination, and cooperation in environmental protection.

The Murmansk speech was followed by immediate negative and positive reactions as well as a series of diplomatic discussions and negotiations to ensure peace and stability in the region and to increase functional cooperation. Due to long-range and regional pollution, in particular radioactivity, concern for the environment resurged in the 1980s as the subject of a number of different threats in the Barents Region, as well as the entire Arctic (e.g., Heininen and Lomagin 2017). Mostly emanating from the Russian side of the Barents Sea, nuclear safety and air pollution became a significant concern and threat, particularly to Norway.[10] As a result, the environment became a new security issue in the region and the main target of a new international cooperative framework, negotiation for the adoption of the Arctic Environmental Protection Strategy (AEPS), began in 1989. It culminated in 1991 with the adoption in Rovaniemi of a common document signed by the governments of the eight Arctic states and Arctic indigenous peoples' organisations. This non-binding agreement was the first major political achievement after the Cold War and marked the introduction of an Arctic dimension into Northern European politics.

Building on the AEPS and need to enhance collaboration at the regional level, new governance structures were also created to support this cooperation including the Barents Euro-Arctic Council (BEAC) established in 1993 and the Arctic Council established in 1996. The first step towards the creation of the Arctic Council occurred in the late 1980s and was accelerated in 1991 after the AEPS was signed. After several years of meeting, the 1996 Ottawa Declaration formally established the Arctic Council as an inter-governmental forum for promoting cooperation among the Arctic States – Canada, the United States, Russia, Denmark, Iceland, Sweden, Norway and Finland – with the involvement of the Arctic Indigenous communities, but excluding representations of other Arctic inhabitants and sub-national governments. The mandate of the Arctic Council focuses essentially on increasing cooperation in the domain of environmental protection and sustainable development, leaving military issues aside as too sensitive issues.

The creation of these new regional structures set the stage for defining the Arctic region and the Barents Region as distinctive, international political regions (Young 1996; Käkönen 1996; Artic Council 2004; Griffiths 1988; Heininen 2009). This was also meant to stabilise the post-Cold War Arctic and institutionalise interstate relations in the Arctic region, as well as develop a new governance framework for reinforcing functional cooperation among Arctic countries, regions and sub-regions.

4 The institutionalisation of the Barents Region's cooperation

In December 1991, the Soviet Union finally collapsed, and the international geopolitical landscape changed. With it came the opportunity for the Nordic countries to reassess their foreign policies to work with, rather than against, Russia with the ultimate aim of decreasing military tension from the Cold War period and increasing political stability in the European North, as well as the entire Arctic Region (e.g. Heininen 2009). The Norwegian Foreign Minister, Thorvald Stoltenberg, first proposed the concept of regional cooperation to his Russian counterpart, Andrei Kozyrev as early as March 1992, a mere three months after the Soviet Union was formally dissolved (Stokke and Tunander 1994, 1). His proposal was based on the experience of the Baltic Sea cooperation and the Council of Baltic Sea States (CBSS). In April 1992, the governors of the Murmansk and Arkhangelsk provinces in Russia were invited to a meeting in Tromsø to discuss possible cooperation. In January 1993, a joint conference of the Foreign Ministers of Norway, Sweden, Finland and Russia took place in Kirkenes just beside the Norwegian-Russian border. It was at this ministerial meeting that the Barents Euro-Arctic Region, as well as intergovernmental and interregional cooperation on the region, was formally established.[11]

The Kirkenes Declaration was signed in January 1993 in Kirkenes, and set out the objectives and purposes of the cooperation. It listed 'environmental protection' in line with the Arctic Environmental Protection Strategy and 'sustainable development' in line with the Rio Declaration on Environment and Development, as the main objectives of the collaboration. Further, while participants of the cooperation indicated 'that the environmental dimension must be fully integrated into all activities in the Region', other areas of cooperation are also listed in the document, such as economic cooperation, science and technology, regional infrastructure, tourism, education and cultural exchange.

In relation to economic cooperation, the Kirkenes Declaration also makes clear that the environment should be preserved and that 'the principles of environmental soundness and sustainability in all fields of economic co-operation' should be observed. Finally, the cooperation specifically targets the situation of indigenous peoples, notably in support of the restoration and preservation of Nenets and Sami cultures, and stresses that 'wider human contacts and increased cultural co-operation in the Region should be encouraged to promote constructive co-operation and good neighbourly relations'. Overall, cross-border relations, with an aim to increase mutual confidence, are at the centre of the cooperation.

From a functional perspective, the Barents cooperation operates at different levels, on a dual level structure, which involves both national and regional governments.

Firstly, at the national level, there is the Barents Euro-Arctic Council (BEAC), an intergovernmental forum with the purpose 'to serve as a forum for cooperation among the participants'.[12] The BEAC is made up of the Foreign Ministers of member countries, Denmark Finland, Iceland, Norway, Russia, Sweden and the European Union's Commission. There are also nine countries with observer status: Canada, France, Germany, Italy, Japan, the Netherlands, Poland, the United Kingdom and the United States of America.[13] The chairmanship of the Council rotates between Finland, Norway, Russia and Sweden with each country holding the chair for two years at a time.[14] For example, for the period 2015–2017 the chairmanship was being held by Russia, and for 2017–2019 it is being held by Sweden.[15] Initially the BEAC met annually, but in recent years, it meets biennially. Between meetings, the Committee of Senior Officials deals with administrative matters. This committee is made up of civil servants from the member states and from the European Commission, and meets regularly to consider progress and to establish working groups on various topics relevant to the cooperation.[16]

Second, at the regional level, there is the Barents Regional Council (BRC) as a parallel body for interregional cooperation. While the BRC initially included seven member regions, today it comprises 14 member counties and a representative of the indigenous peoples in the northernmost parts of Finland, Norway and Sweden and northwest Russia.[17] The most recent member to join was the Finnish county of North Karelia, which was approved as a member in November 2016, having held observer status for many years. There are also two organisations with observer status: The Council of Christian Churches in the Barents Region and The Parliamentary Association of North West Russia.[18] With the same purposes as the BEAC, the BRC has a mandate to support and promote cooperation and development in the Barents Region. The purpose of the BRC is also to recognise and share local knowledge and to provide the opportunity to identify and work on cooperative projects. The chairmanship of the Regional Council rotates between the countries in Norway, Sweden, Russia and Finland every two years with the provision that the same country cannot chair both the interregional council and the international council at the same time.[19] The executive body of the BRC is the Barents Regional Committee, which is composed of civil servants from the member counties.

Furthermore, 15 working groups have been established to enhance cooperation on issues relevant to the Barents Region. Each working group is able to enhance cooperation in its area of expertise and coordinates projects across the Barents Region. Some of the working groups, such as the Working Group on Environment and the Joint Committee on Rescue Cooperation, are working groups established by the BEAC. Correspondingly, the BRC has established others, such as the Regional Committee on Environment and the Regional Working Group on Transport and Logistics. There are also a number of joint working groups, including the Joint Working Group on Energy and the Joint Working Group on Youth.

As part of these, the Working Group of Indigenous Peoples (WGIP) established in 1995 on a permanent basis. It is composed of representatives of the Sami, the Nenets and the Veps of the Barents Region. On a daily basis, the main goal of the WGIP is 'to secure indigenous peoples' rights, foundation for trade, society, culture and language through implementation of the Action Plan of Indigenous Peoples' (WGIP 2017, 7), which constitutes WGIP's policy development framework.[20] The 2017 Action Plan lists different issues and a set of measures to implement in order to secure indigenous peoples' rights.[21] With this action plan, the WGIP sets the priorities of indigenous peoples in the region and can advise the BEAC or the BRC when they initiate projects concerning them. Unlike other regional working groups, in addition to its operational role, the WGIP has also an advisory role to both the BEAC and the BRC, which consequently gives it a political dimension. The chair of the WGIP sits also as a member of the Committee of Senior Officials in the BEAC and the Barents Regional Committee, which gives the WGIP a permanent right to attend all meetings organised by the councils. Thus, WGIP constitutes an important platform for indigenous organisations to strengthen their political stance at the regional level, as well as to enhance their cooperation with the national, regional and local authorities of the Barents Region's countries and counties (WGIP 2017, 7).

5 The development of the Barents cooperation, its achievements and challenges

Those who established the Barents Euro-Arctic Council and signed the 1993 Kirkenes Declaration hoped that cooperation would lead to stability, progress, peace and security throughout the region and eventually promote the wellbeing of the entire population. To celebrate its twentieth anniversary, the Barents countries adopted a new declaration at the Barents Summit in 2013. In the document (Kirkenes Declaration 2013) all states reaffirmed their commitments to the cooperation, which is justified by 'the important role the Barents cooperation has played in strengthening mutual trust, stability and security in Europe, by joint efforts in northern Europe based on the shared commitment to indivisible and comprehensive security'. Twenty years after its establishment, these renewed statements certainly demonstrate that the cooperation is still playing an important role in the enhancement of security and cooperation in the region. In fact, as an outcome of this development, the Barents Region is also no longer a periphery of Europe, as it was before 1993, and 'the demand for cooperation is greater today than ever' (Kirkenes Declaration 2013).

During these 20 years the geopolitical situation of the Barents Region, as well as the entire Arctic, has significantly changed from confrontation during the Cold War period, when the Barents Sea area was mostly a military 'theater', to an international, much more functional, cooperation (see Heininen and Lomagin 2017, 269–274). In the 2010s there are two main developments or tendencies. On the one hand, the Arctic region, with high geopolitical stability, is globalised and impacted by grand environmental challenges, in particular (rapid) climate change, and interests and plans for mass-scale exploitation of natural resources located in the region. On the other hand, there is geopolitical tension between Russia and the West (in particular the USA and the EU) much due to the Russia's annexation of Crimea and the Ukrainian crisis/war.

However, despite this tension and colder political climate between the Arctic states, the Barents Region's official and unofficial cooperation remains largely unaffected. Behind this lies Arctic 'exceptionalism', as high geopolitical stability in the Arctic remains and is resilient (see Heininen 2016). In fact, it has been suggested that 'the Arctic to this point remained largely insulated from wider geopolitical issues following 2014 – and in some cases cooperation has deepened' (Clifford 2017) and contacts between regional and local representatives still goes on (Nilsen 2016a). Compared to the Baltic Sea Region, it was also stressed that in the Barents Region, 'it is not as high tension' and that there is 'good cooperation with Russia on a lot of common areas and issues of importance' (Nilsen 2016b). As a result, the Barents regional cooperation stands strong (at the moment), and even in this difficult geopolitical context, it continues to promote the development of the region across national borders.

Regarding its achievement, over the last twenty years, the cooperation in the Barents Region has led to the sharing of expertise, technology, finances and other resources which has resulted in a number of improvements to the environment for the benefit of the whole of the region and, as a result, to the security of the people in the region. Although the environmental threats have not yet been completely eradicated (see Chapter 2.1), thanks to the cooperation the threat of major environmental disasters, with all the implications for the environment and for human security, has been dramatically reduced. There have also been a number of cultural, educational and other projects with an aim to bring the people(s) of the Barents Region together, some organised by the formal institutions of the Barents Region, others merely as a result of increased community cooperation. These include exhibitions and festivals, such as the Barents Spektakel 2016, an impressive display of art, film, workshops, seminars, music and sport in the nearby towns of Kirkenes (Norway) and Nikel (Russia); and the Calotte Academy, an annual academic gathering and travelling symposium for scientific work and open discussion to implement the interplay between science and politics (e.g. Final Report of the 2016 Calotte Academy, Huotari et al. 2016). Other projects include student exchanges, a multi-stage ski race dubbed the 'Tour de Barents' and a health programme specialising in tuberculosis and HIV/AIDS (e.g. Wilson Rowe 2009, 35–52). In 2016, the Ministers of Culture of the member states announced funding for a number of scholarships across the region to foster cultural cooperation.

However, there are, also weaknesses, uncertainties and criticism regarding the Barents cooperation. These notably relate to the fact that cooperation is neither deep enough nor is there a common historical-cultural identity. As Zimmerbaur (2012) argues 'the idea of [the] region as an imagined community is stronger elsewhere than *in situ*'. It can also be questioned how much activities and tasks are currently performed under the auspices of each working groups. The development of new projects to further cooperation is highly dependent on resources and national support, which are manifestly insufficient to support cooperation. Currently, 'the majority of Barents Region consider the current level of financing insufficient for their joint activities' and to finance every day work (BEAC 2015, 7). The main issue in financing the Barents cooperation relates to a 'lack of financial mechanisms that would encompass the whole Barents Region'. This situation 'prevents regional actors from developing geographically-wide projects and leads to unequal access to financial opportunities of different regions' (BEAC 2015, 43). In addition, non-governmental organisations and other civil society actors have also indicated that financial support remains insufficient in supporting small-scale activities (BEAC 2015, 51). Thus, it seems that cooperation could be improved if more funding was made available.

Furthermore, while the working groups and programme developed under the auspices of the Barents cooperation have obtained tangible results in reinforcing cross borders ties and relations, there are still issues that continue to challenge the wellbeing of the Barents population that have not been addressed institutionally. As already mentioned, this includes the grand scale of environmental challenges, in particular the impact of global warming and the development of natural resources located in the region. In addition, whereas the Ukrainian crisis has not affected the Barents cooperation directly, it has strained the economic development of the region and increased political tensions between states, a topic that is largely outside the purview of the Barents cooperation. Thus, any real assessment of the Barents situation cannot be fully examined through the lens of the existing cooperation framework. To this degree, there is an entire human experience that exists beyond the cooperation institutions.

Therefore, while the wellbeing of peoples of the region constitutes the overriding goal of the Barents cooperation, and collaboration has enabled better cross-border relations over the years, there is a continuous need to consolidate cooperation in addressing present and emerging societal challenges. Against this background, the following chapters intend to assess the challenges faced by the Barents population in relation to several human security issues in order to assess how regional cooperation can address the populations' interests and wellbeing.

Conclusion

Since its establishment, the vision of the Barents Region's cooperation has been 'to improve peoples' living conditions, to encourage sustainable social and economic development, and to have a peaceful and sustainable development in the northernmost part of Europe' (BRC 2014, 6). The Kirkenes Declaration has been the basis for the establishment of a comprehensive framework to ensure functional cooperation in many fields, excluding military-security, such as the environment, economy, human health, tourism and cultural interaction.

Despite occasional tensions between the governments of Russia and the Nordic countries, the contacts between sub-national governments of, and regional capitals in, the Barents Region have remained, and contacts between peoples and civil societies continue to be supported through the institutional mechanisms that have been created for enhancing cooperation in the region (e.g., Nilsen 2016a). This precisely accords to the ultimate aim of region-building with states as major actors – one of the main trends of the post-Cold War Arctic IR and geopolitics (Artic Council 2004) – i.e. the establishment of the BEAR as an international, cooperative region located in the former military theatre of the Cold War. Therefore, it is possible to conclude that from the point of view of the states, the main objective has been completed, even though other objectives, such as sustainable development require more efforts (Heininen 2009).

In effect, the Barents cooperation continues to operate as an institutional forum to promote dialogue and concrete cooperation with the purpose of strengthening regional stability, sustainable development and the wellbeing of the Barents population. The Barents cooperation is truly an achievement, especially considering the historic context of the Cold War. The borders of the socialist Soviet Union were closed through much of the 20th century, and there was very little contact or cooperation across the national borders of the countries surrounding the Barents Sea, and with the open, democratic Nordic countries in the West. Such an achievement highlights the deep value of, and commitment to, mutually beneficial cooperation across borders. Indeed, for centuries the populations of the Barents Sea area have been closely intertwined through trade and cultural exchanges, as the Pomor Trade era well shows. Behind this, the peoples of the region share several features that were similar, especially for indigenous peoples and settlers, whose families have lived there for centuries. A problem of the past was that there were few, if any, opportunities for those living on either side of the border to meet or to share culture or expertise during most of the 20th century. Despite this, it is evident that the population has established strong relations under the auspices of the Barents cooperation.

As argued, one of the key motives for cooperation between the countries in the Barents Region was the desire to ensure the wellbeing of the population across the borders that separate them by consolidating cooperation in key areas of issues relevant to the region. Despite criticism that the interregional cooperation is not deep enough, or that the region lacks cultural identity or financial resources, the work of the two Barents Councils shows there is an ever-growing amount of cooperation among the countries of, and counties in, the Barents Region. This ranges across diverse sectors such as youth, culture and human health to tourism and transport, and has strengthened the societal bonds of the Barents Region's populations. The sharing of ideas, expertise and resources has led to improvements in standards of living across the region. Alongside the exchange of culture and ideas, higher living standards and increased economic opportunities for all ensures greater levels of economic and political stability which promotes human security.

Every time a project results in the collaboration between previously opponent countries, it leads to greater levels of understanding and cooperation. This form of cooperation is cumulative, and contributes to increased levels of peace and security within the region. While the Barents cooperation certainly faces a number of challenges, it remains an important framework to enhance human security among its civil societies and to strengthen the societal bonds of its peoples. Indeed, the Barents Region is a society with shared human security concerns.

Notes

1 For more information about the Barents Region see Olsson et al. (2016), *The Encyclopedia of The Barents Region*.
2 www.barentsinfo.org/Barents-region/Facts (visited on 6 October 2016).
3 Declaration of Cooperation in the Barents Euro-Arctic Region, Conference of Foreign Ministers in Kirkenes, 11 January 1993.
4 www.barentsinfo.org/Barents-region/Facts visited on (6 October 2016).
5 Webpage of the Barents Council: www.barentsinfo.org/Barents-region/Facts (visited on 7 March 2017).
6 In Norway, nearly 100% of the electricity demand is produced locally with hydropower in the counties of Nordland, Troms and Finnmark. Since the beginning of the 21st century, Sweden is also investing in major wind power development projects, especially in the counties of Norrbotten and Vasterbotten. While Finland is also following a similar trend, Russia lags behind in term of investment concerning renewable energy.
7 For more detailed information about the history of the Barents see Elenius et al. (2015) and Olsson et al. (2016).
8 At that time, King Magnus Karl Eriksson jointly ruled the territory of Norway and Sweden but the kingdom remains politically distinct.
9 Webpage of the Nordic Co-operation: www.norden.org/en/fakta-om-norden-1/history-of-the-nordic-region/five-welfare-states-in-a-global-world-1920 (visited on 7 March 2017).
10 By the time the Soviet Union ended, northwest Russia was the most highly nuclearized region in the entire world. The city of Murmansk was home to the Russian Northern Fleet, which controlled large numbers of nuclear submarines, and there

were estimated to be at least 270 nuclear installations, both military and civil situated in the region. There had been very little investment in the region and little attention paid to quality or to safety during the final years of the USSR, which meant that the nuclear installations posed a huge environmental threat to the entire region.

11 Declaration Cooperation in the Barents Euro-Arctic Region Conference of Foreign Ministers in Kirkenes, 1993.

12 Annex to the Kirkenes Declaration Cooperation in the Barents Euro-Arctic Region Conference of Foreign Ministers Kirkenes, Norway, 11 January 1993 Terms of Reference for the Council of the Barents Euro-Arctic Region, art 1.

13 Administrative Manual for the Barents Euro-Arctic Council, 17 June 2008.

14 Administrative Manual for the Barents Euro-Arctic Council, 17 June 2008.

15 Webpage of the Barents Council: www.barentscooperation.org/en/Barents-Euro-Arctic-Council/Chairmanship (visited on 4 March 2017).

16 Administrative Manual for the Barents Euro-Arctic Council, June 17 2008

17 The Barents Region includes the following counties or their equivalents:

- in Finland: Kainuu, Lapland, Oulu Region and North Karelia.
- in Norway: Finnmark, Nordland and Troms
- in Russia: Arkhangelsk, Karelia, Komi, Murmansk and Nenets.
- in Sweden: Norrbotten and Västerbotten.

18 Barents Regional Council, Barents Regional Committee Terms of Reference Adopted by the Barents Regional Council on November 14 2012 in Oslo, Norway.

19 Barents Regional Council, Barents Regional Committee Terms of Reference Adopted by the Barents Regional Council on November 14th 2012 in Oslo, Norway.

20 Ibid.

21 Ibid.

References

Artic Council. 2004. *Arctic Human Development Report*. Edited by N. Einarsson, O. Young and A. Nilsson. Akureyri: Stefansson Arctic Institute.

Barents Info. 2017. Veps. www.barentsinfo.org/Contents/Indigenous-people/Veps

BEAC. 2015. *Financing of the Barents Cooperation, Report of the BEAC Ad Hoc Working Group on Financial Mechanism Study*. Helskini: Ministry for Foreign Affairs of Finland.

BRC. 2014. The Barents 2014–2018 programme. www.barentsinfo.fi/beac/docs/Barents_Programme_2014_2018_Brochure.pdf

Carrillo, J.L.M. 1998. *Regional Security Building in Europe. The Barents Euro-Arctic Region*. Umeå: CERUM.

Clifford, R. 2017. How has cooperation in the Arctic survived Western-Russian geopolitical tension?http://pagebuilder.arctic.arcpublishing.com/pb/voices/analysis/2017/01/04/how-has-cooperation-in-the-arctic-survived-western-russian-geopolitical-tension/

Elenius, L., Tjelmeland, H., Lähteenmäki, M., GolubevA.*et al.* (Eds). 2015. *The Barents Region: A Transnational History of Subarctic Northern Europe*. Oslo: Pax Forlag AS.

Godal, B.T. 1995. *In Nansen Footsteps. The Barents Cooperation: A Vision for a Better Europe*. London: Royal Geographical Society.

Griffiths, F. 1988. Introduction: the Arctic as an international political region. In *The Arctic Challenge: Nordic and Canadian Approaches to Security and Cooperation in an Emerging International Region*. Edited by K. Möttölä. Boulder: Westview, 1–14.

Gyllenhaal, L. 2017. Conflicts in the Barents Region. www.barentsinfo.org/Contents/History/Conflicts-in-the-Barents-region

Hacquebord, L. 1995. In search of Het Behouden Huys: a survey of the remains of the house of Willem Barentsz on Novaya Zemlya. *Arctic* 48: 248–256.

Heininen, L. 1999a. The Barents Region. In *Nordic Region-Building in a European Perspective*, edited by H. Baldersheim and K. Ståhlberg. Ashgate. Athenaeum Press Ltd.

Heininen, L. 1999b. Euroopan pohjoinen 1990-luvulla. Moniulotteisten ja ristiriitaisten intressien alue (The European North in the 1990s – a region of multifunctional and conflicting interests). *Acta Universitatis Lapponiensis 21 – Arctic Centre Reports* 30. Rovaniemi.

Heininen, L. 2009. The Barents Region in the state interests and international politics. *Barents Journal* 1(7): 5–10.

Heininen, L. 2016. High Arctic stability as an asset for storms of international politics – an introduction. *In Future Security of the Global Arctic. State Policy, Economic Security and Climate*, edited by L. Heininen. Basingstoke: Palgrave Macmillan.

Heininen, L. and Lomagin, N. 2017. Geopolitics, security, and globalization. In *The Encyclopedia of The Barents Region* (1,2), edited by M.O. Olssonet al. Oslo: PAX FORLAG.

Huotari, J., Olsen, L. and Zojer, G. (Eds). 2016. *Final Report of the 2016 Calotte Academy: Resilience Related to Sustainable Development in Globalization*. May 30–June 5, 2016. Finland, Russia, Norway. https://calotte-academy.com/sites/default/files/2017-08/CA-2016-Final_Report.pdf.

Käkönen, J. 1996. Dreaming the Barents Region. Interpreting cooperation in the Euro-Arctic Rim. Tampere Peace Research Institute, *Research Report* No. 73.

Kolga, M. 2001. *The Red Book of the Peoples of the Russian Empire*. Tallin: NGO Red Book.

Nilsen, T. 2016a. Barents cooperation in winds of change. *Arctic Yearbook*. www.arcticyearbook.com/commentaries2014/127-barents-cooperation-in-winds-of-change

Nilsen, T. 2016b. United against Russia's aggression. *Independent Barents Observer*. https://thebarentsobserver.com/en/2016/11/united-against-russias-aggression.

Olsson, M.O.et al. (Eds) 2016. *Encyclopedia of The Barents Region* (1, 2). Oslo: Pax Forlag.

Pravda1987. Mikhail Gorbachev's Speech in Murmansk, October 15, 1987.

Stokke, O.S. and Tunander, O. (Eds). 1994. *The Barents Region: Cooperation in Arctic Europe*. International Peace Research Institute. London, Thousand Oaks, New Delhi: Sage.

Stoltenberg, T. 1992. The Barents Region: reorganizing Northern Europe. *International Challenges* 12(4): 5–12.

Tuisku, T. 2004. Nenets. http://www.barentsinfo.org/Contents/Indigenous-people/Nenets.

Wilson Rowe, E. 2009. *Russia and the North*. Ottawa: University of Ottawa Press, 35–52.

Working Group of Indigenous Peoples in the Barents Region. 2017. *Action Plan for Indigenous Peoples in the Barents Euro-Arctic Region 2016–2018*. https://www.barentsinfo.fi/beac/docs/ActionPlan_2017-2018_ENG.pdf

Young, O. R. 1996. *The Arctic Council: Marking a New Era in International Relations*. New York: The Twentieth Century Fund Report.

Zimmerbauer, K. 2012. Unusual regionalism in Northern Europe: the Barents Region in the making. *Regional Studies* 47: 89–103.

Part II
Assessment

2.1 Environmental security in the Barents Region

Sarah Mackie

Environmental security is vital to the future of the Barents Region and its population, which relies on the environment for their food and water, their health and their livelihoods. When the environment is damaged or harmed, the security of human communities is threatened.

1 Definition

The idea that environmental security is a vital part of human security is not new. In 1994 the United Nations Development Programme published the Human Development Report which introduced the concept of human security linked to factors other than conflict (UN Development Programme 1994). Right from the very beginning, environmental security was included on the list of seven categories of threats to human security (UN Development Programme 1994).

Despite this, there are few formal definitions of environmental security. The Millennium Project provided an overview of the definitions which existed at the turn of the century. The report found no single agreed definition but the authors created a synthesis definition which has since been widely used (Glenn et al. 1997; Coskun et al. 2008; Hull et al. 2009). They defined environmental security as 'environmental viability for life support with three sub-elements: (1) preventing or repairing military damage to the environment; (2) preventing or responding to environmentally caused conflicts; and (3) protecting the environment due to its inherent moral value' (Glenn et al. 1997). While this definition covers much of what is meant by environmental security, it is not an exhaustive definition as it does not consider the requirement of environmental security for food, health and livelihoods.

In 1994, the Russian Inter-Agency Commission on Environmental Security adopted the following definition:

> Environmental security is protectedness of natural environment and vital interests of citizens, society, the state from internal and external impacts, adverse processes and trends in development that threaten human health, biodiversity and sustainable functioning of ecosystems, and survival of

> humankind. Environmental security is an integral part of Russia's national security.
>
> (Glenn et al. 1997; Russian Security Council 1996)

From a purely environmental point of view it includes biodiversity and sustainable ecosystems and from a human security perspective it covers both health and survival. It fails to consider livelihoods but is a broader and more complete definition than the synthesis definition derived from the Millennium Project (Glenn et al. 1997).

Taking the positive aspects of the definitions discussed above, this analysis will work on the basis that environmental security means the protection of the natural environment to ensure the sustainable functioning of ecosystems and biodiversity both for the intrinsic value of nature and for the support of the survival, health and livelihoods of both individuals and communities. Communities which are environmentally secure will be less vulnerable to, and will be able to survive, changes in their environmental surroundings.

2 Contextualisation

The environment of the Barents Region is remarkably diverse, covering as it does Europe's Arctic region, parts of Russia's Arctic and areas of land which lie south of the Arctic Circle (Barents Info 2017). The Barents Euro-Arctic Council described the region as 'compris[ing] most of the remaining pristine wilderness areas in Europe. It has unique values of biodiversity, natural resources and high recreational qualities' (Barents Euro-Arctic Council 1994). The region is wild and remote, icy, raw and often severe but is, at the same time, exceptionally beautiful.

The southern part of the Barents Region is characterised by coniferous forests while the north is predominantly unvegetated Arctic tundra in the east and mountains in the west (Barents Info 2017; Olson et al. 2001). While the climate of the Barents Region is cold, it is not the coldest part of the Arctic; very little of the area falls within the 10°C isotherm in July (Przybylak 2015, 1–4). Winter temperatures are lowest in the east where the average temperature in January can reach −20°C and are warmest on the Norwegian coastline which averages only −3°C in January (Rekacewicz 2005a).

The Barents Region is home to 5.3 million people (Barents Euro-Arctic Council 2017). Around 7% of the population is reliant on the natural environment for their livelihood, undertaking activities including hunting, fishing, reindeer herding and forestry (Plummer and Baird 2013). Other people supplement their diet with food such as berries and mushrooms foraged from their surroundings (Kozlov and Barcan 2000, 514, 516). For the indigenous and non-indigenous populations, the environment is vital for both cultural and economic reasons. Reindeer herding has a deeply cultural aspect to it with the practice dating back thousands of years (Stammler 2005; UNESCO 2006). Other activities, such as berry picking, also have a significant cultural element (Fryer et al. 2010, 16). A large, and ever increasing, part of the economy of the Barents Region is based on tourism with the pristine wilderness being one of the region's major attractions (AMAP 2017a, 12; Stonehouse and Snyder 2010).

3 Assessment of environmental security in the Barents Region

The Barents Region faces some severe environmental threats, in particular from climate change, industrial pollution and the risk posed by nuclear contamination.

3.1 Climate change

There is little doubt that climate change is one of the most pressing environmental threats currently facing the Arctic, both in the Barents Region and beyond. There is now scientific consensus that climate change across the planet is occurring and that it is caused by human activity (Pachauri et al. 2014). The Intergovernmental Panel on Climate Change ('the IPCC') reported in 2014 that '[w]arming of the climate system is unequivocal, and since the 1950s, many of the observed changes are unprecedented over decades to millennia' (Pachauri et al. 2014, 40).

While climate change is a global problem, temperatures in the Arctic are rising almost twice as fast as temperatures in other parts of the world and, in the foreseeable future, the impact of climate change is predicted to be greater in the Arctic than in any other part of the planet (Solomon et al. 2007, chap. 4; AMAP 2005). In 2011 the Snow, Water, Ice, Permafrost in the Arctic Report (SWIPA) reported that substantial changes were being seen in the Arctic environment (AMAP 2011, 22). The changes were so great that the report concluded that earlier predictions made by the IPCC in 2007 had underestimated the likely rate of change, particularly in sea ice measurements (IPCC 2007; AMAP 2011, chap. 9). The findings of SWIPA 2017 have made it clear that changes in the Arctic are becoming increasingly dramatic. Temperatures in the Arctic in recent years have been unprecedented, January 2016 being a full 5°C warmer than the average from 1981–2010 (AMAP 2017b, 10). According to the USA's National Snow and Ice Data Centre, seven of the months in 2016 showed the lowest sea ice extent since records began 38 years ago (National Snow and Ice Data Center 2016). SWIPA 2011 reported that temperatures in the Arctic's permafrost had risen by up to 2°C causing the southern limit of the permafrost to move north (AMAP 2011, chap. 2). By 2017 even these concerning numbers had been exceeded, with permafrost in the high Arctic 0.5°C warmer than had been recorded as recently as 2007–2008 (AMAP 2017b, 66). It is now predicted that the Arctic Ocean will be ice free in the summer within two decades (AMAP 2017b, viii). Research shows that it is likely that the Barents Sea will be the first Arctic sea to be free of ice, not just in the summer but for the entire year by the middle of the century (AMAP 2017a, 9). This will impact marine life, sea levels and weather patterns (AMAP 2017a, 9). On shore, the increasing temperatures will see the northward movement of 'species, pests and diseases' which will have a dramatic effect on the food chain and the human societies (AMAP 2017a, 9).

As the Arctic warms, species which are adapted to the cold environment and are unable to cope in a warmer climate will become extinct; warming temperatures will also result in new predators moving into the Arctic areas with a resultant risk to the species which will become prey (AMAP 2011). Species already noted to be at risk include a number of mosses and lichens and some small herbivores such as lemmings and voles which will also affect predators further up the food chain reliant on these species such as reindeer (Larson et al. 2014, 1580). The vegetation in the area is changing with more abundant deciduous shrubs and grasses being observed growing north of the Arctic Circle (Larson et al. 2014, 1578–80). The tree line is moving steadily north, displacing tundra in favour of boreal forest (Larson et al. 2014, 1570, 1580; AMAP 2011, chap. 12). The permafrost is melting, causing the land to heave and then collapse as it melts and refreezes, changing the biochemical make up of water in rivers and lakes and allowing greenhouse gases trapped in the ice to be released, thereby adding to the overall warming effect (Collins et al. 2013, 1096; Larson et al. 2014, 1573, 1586). In almost every aspect of the environment in the Barents Region changes are being seen and the rate of change is likely to increase over the coming years.

There is a resulting threat to human security in the Barents Region and, in particular, in the Arctic parts of the region. This is partly because of the speed at which change is occurring and partly because the types of communities found in the Barents Region (predominantly isolated, indigenous communities) have fewer adaptive choices than those located elsewhere (Larson et al. 2014, 1593–1596). These communities are well adapted to life in the cold and their lifestyles have developed to allow them to thrive even in the harshest of winters. As the temperatures rise, however, they are at risk of flooding from melting ice and snow, have fewer winter travelling opportunities and are at increased risk of injury when they venture out on ice. Melting permafrost will cause damage to vital community buildings in remote villages such as clinics and schools and will threaten human health though the destruction of village sanitation and food storage systems (Butler 2016). There is even a risk of disease from the melting of previously frozen corpses as was recently experienced by the Nenets people in Russia (Larson et al. 2014, 1582–1590; Siberian Times 2016). The Sami and Nenets peoples who are reliant on reindeer herding to feed their families and to provide income through the sale of skins and meat have reported that increasing amounts of winter rain which, unlike snow, freezes on contact with the ground is making it difficult for reindeer herders to find grazing areas which are not coated in ice (Larson et al. 2014, 1581–1584; AMAP 2011, chap. 10; Labba 2015).

There is an urgent need to work to prevent future changes in the region and to ensure that any changes which do occur are kept to a minimum. While much of the warming experienced in the Arctic is caused by global greenhouse gas emissions, the effect of global warming is amplified in the Arctic by the existence of surface black carbon deposits (Arctic Council 2013; AMAP 2015). Black carbon is an atmospheric particle pollutant which, if it falls on ice or snow, creates a black or dark patch on the ground which absorbs more heat than the ice or snow would, further exacerbating the warming of the Arctic (Arctic Council 2013; AMAP 2015). This particular effect of black carbon is fairly localised because it falls to earth close to where it is emitted so carbon produced in or near the Arctic has a greater impact on the Arctic than carbon emitted elsewhere (Arctic Council 2013; AMAP 2015). Sand et al. discovered that black carbon emitted in the Arctic has a warming effect that is five times that of black carbon emissions from lower latitude areas and concluded that black carbon deposits on snow account for two-thirds of the increase in surface temperatures in the Arctic (Sand et al. 2013; AMAP 2015, 16).

One of the biggest contributors of black carbon in the Arctic is gas flaring (Stohl et al. 2013) which is the name given to the practice of burning off unwanted hydrocarbon gases in the oil and gas industries. It is estimated that gas flaring accounts for 66% of the black carbon emitted above 66°N annually and almost 80% of that emitted above 60°N during the winter when there are fewer forest fires (Stohl et al. 2013, 8836). This is in stark contrast to the global figures where gas flaring accounts for a mere 3% of the total black carbon emissions (Stohl et al. 2013, 8836). Almost all of the high latitude gas flaring takes place in the Barents Region and eastern Russia (Stohl et al. 2013, 8836). While black carbon emissions may not be the first priority in the worldwide fight against climate change, the particular nature of the impact on the Arctic and the way in which it causes an increase in temperature in the area in which it is emitted means that it should be a priority for the Barents Region.

Alongside prevention of climate change, there is also a need to consider how people and communities adapt to the changing circumstances as it is unlikely that the changes being brought about by climate change can or will be halted entirely. While indigenous communities have adapted to change in the past, the fear is that the rate of change expected as a result of climate change, combined with a lifestyle which is far more 'circumscribed' by political, social and economic situations than it was in the past, will make adaptation difficult (Nuttall 2007). Plans need to be put in place in order to secure the environment both for its own sake and for those who rely on it; plans made and implemented now may limit the temperature rises and also enable mitigation and adaptation to occur before the situation in the Barents Region reaches crisis point, preventing the human and environmental tragedies which could be precipitated by climate change.

3.2 Industrial pollution

The environment of the Barents Region is also threatened by the pollution that is emitted into the air and water sources by industrial sites situated within the region. As a result of the proximity of the industrial areas to the borders of the countries which form the Barents Region, much of this pollution is transboundary, causing environmental damage in countries which have no control over the industrial site emitting the pollution.

3.2.1 Russia

The Kola Peninsula is renowned for its two major smelterworks in Nickle and Monchegorsk. These smelterworks have traditionally been two of the largest global emitters of sulphur dioxide (Kashulina et al. 1997). Acid rain, caused by sulphur dioxide emissions, has had a large environmental impact on the area surrounding the smelterworks, both in Russia and across the borders in Norway and Finland, with environmental degradation, including the destruction of moss cover, damage to trees and forests, the death of fish and soil erosion, widely reported (AMAP 2006; Kashulina et al. 1997; Norseth 1994; Rekacewicz 2005b; Tuovinen et al. 1993). In 1991, the area surrounding the smelterworks was dubbed 'one of the world's worst ecological situations' and a 'technogenic wasteland' because of the impact of acid rain on the local environment (Kashulina et al. 1997; Kryuchkov 1990; Pulkkinen 1991). Both of the smelterworks were designated as 'Environmental Hot Spots' and work has been undertaken to reduce the level of sulphur dioxide emissions (Bambukyak et al. 2013, 25–27). The emission levels at the Nickel smelterworks have been significantly reduced but it remains one of the largest air polluters in the region and the continued levels of air pollution remain a threat to the environment (Bambukyak et al. 2013, 25–26). At the Monchegorsk smelterworks, investment of €122 million has brought the emissions within the Maximum Allowable Emission levels and is leading to the regeneration of flora and fauna in the area (Bambukyak et al. 2013, 26–27).

Many of the industrial sites in north west Russia discharge waste water containing pollution in the form of heavy metals and other contaminants into local streams, rivers and lakes (Bambukyak et al. 2013, 23–24). An example of this is found at the Apatit phosphate ore mine in the Murmansk region which discharges contaminated water into nearby rivers and lakes (Bambukyak et al. 2013, 28–29). Work undertaken since 2003, including the construction of a mine water treatment plant, has led to a reduction in the levels of aluminium and fluorine in the discharged water but the levels of contaminants remain above allowable levels (Bambukyak et al. 2013, 28–29). Pollution of this kind disrupts the delicately balanced ecosystems of the rivers and lakes and because contaminated water carries the pollution downstream the impact can be widespread (Malinovsky et al. 2002). Many of the contaminants being released are toxic to both flora and fauna, either directly or through accumulation over time (Moiseenko and Kudryavtseva 2001). Evidence suggests that both plants and animals are particularly susceptible to the negative impact of toxins during the polar night as a result of both the cold and the lack of light meaning that the effect of the water pollution on the local environment from the industrial sites on the Kola Peninsula is amplified during the winter (Lemly 1996; Moiseenko 1999, 37).

3.2.2 Norway

Industrial pollution is not just a problem in northwest Russia; there have been recent controversies in the Nordic countries as well. In December 2016, the Norwegian government upheld a permitting decision allowing a mining company to open a copper mine in the Nussir and Ulveryggen mountains in the Kvalsund municipality in the far north of Norway (Klima-og miljødepartementet 2016). The permit has been controversial because it will allow the company to dispose of mine tailings in the nearby Reppar Fjord (Staalesen 2016). Norway is one of only three countries in the world which allow mining waste to be put into the sea (Nilsen 2016b). The permit issued in this case will allow up to eight square kilometres of the nearby fjord to be filled with mining waste prompting protests from the local Sami community and environmental groups (Staalesen 2016). The organisations Natur og Ungdom and Naturvernforbundet issued a joint statement appealing the decision which said that they 'fear[ed] that the implementation of the project will be the largest pollution scandal in Norway' and highlighted, *inter alia*, the risk of heavy metal leaching and the impact of this on Sami reindeer herders (Natur og Ungdom and Naturvernforbundet 2015). Despite these protests the Minister for Climate and the Environment upheld the issuing of the permit (Klima-og miljødepartementet 2016). It is not Norway's first permit allowing disposal of mining waste in the Arctic Oceans; in the northeastern part of Norway, another company, Sydvaranger, which mines for iron ore, is allowed to dispose of mining waste in the waters of the Bøkfjord near the town of Kirkenes on the Norwegian Russian border (Nilsen 2014).

3.2.3 Sweden

In Sweden, the Blaiken mine located in Västerbotten county is leaking heavy metals (lead, copper and zinc) into the local environment. Blaiken mine was an open cast gold, and later zinc, mine which was abandoned by its owners, ScanMining, in 2007 (Granqvist 2016). The mine is currently leaking heavy metals into the nearby Lake Storjuktan, destroying the flora and fauna of the lake (Granqvist 2016, 2014). The company, now bankrupt, had set aside three million Swedish krona to pay for clean-up but it is estimated that amelioration will cost the Swedish taxpayer 200 million Swedish krone (Granqvist 2016, 2014). Svärtträsk mine, owned by the same company and located only thirty kilometres from Blaiken mine, is leaking contaminated water into a nearby river (Granqvist 2014).

3.2.4 Finland

In Kainuu region, Finland, the Talvivaara mine has proved to be one of the biggest environmental disasters in Finland. Talvivaara is an open pit mine located in Sotkamo with deposits of uranium, nickel and zinc among other metals. Between 2010 and 2013, the mine suffered from three leaks from its gypsum pools, each discharging uranium, nickel and other heavy metals into the surrounding rivers and lakes (Bedford 2015). In 2012, a large number of dead birds were found near one of the mine's waste ponds and a member of staff was killed as a result of hydrogen sulphide poisoning (Bedford 2015). Noxious dust, fumes and water pollution from the mine are estimated to have spread up to 100km (Bedford 2015). The mining company had acted in breach of their permit and three senior managers of the company as well as the bankrupt company itself were prosecuted for environmental crimes (Mines and Communities 2016).

3.2.5 Impact on human security

When air or water is polluted or when acid rain and other pollutants damage the land and the forests and erode the soil, there is a resulting effect on the health of the people who rely on those resources. Air pollution from the industrial sites in north west Russia has been reported to have caused respiratory problems and increased mortality rates from cardiovascular and circulatory diseases as well as cancer (Nieminen et al. 2013; Norseth 1994, 106–107; Pasanen et al. 2012; Revich 1995). Where water sources which are used for drinking water are polluted, the local community must either find alternative drinking water sources or drink the polluted water and risk any potential health consequences. Pollution in water can enter the food chain causing health problems to those who fish in the lakes and rivers which have been polluted. For example, the heavy metals that are contaminating waters in the Kola Peninsula have been linked with human health risks such as neurological defects, cancer, anaemia and damage to bone tissue (Förstner and Wittmann 2012, 2, 27; Moiseenko and Kudryavtseva 2001). Those who use the forests and the land for food are also put at risk of illness as a result of environmental degradation: high levels of contaminants have been found in potatoes, vegetables, berries and mushrooms which are grown or collected, and then consumed, by the communities close to the smelterworks on the Kola Peninsula (Kozlov and Barcan 2000, 514–516). Numerous health problems have been reported in the local population related to eating contaminated food (Kozlov and Barcan 2000, 516).

3.3 Nuclear threat

At the time that the Barents Region was first created, northwest Russia was the most heavily nuclearised part of the world, with large numbers of both military and civilian industrial nuclear reactors as well as many nuclear submarines which formed part of the Russian Northern Fleet (Stokke 1994). Current and historic nuclear activity in the region, both in Russia and in Finland, threatens the environment through the destruction or mutation of the cells of biological organisms resulting in death, radiation sickness, cancer and genetic damage in humans, animals and plantlife (US Office of Technology Assessment 1995, chap. 3).

3.3.1 Dumping of radioactive waste

Despite the levels of nuclear activity in the east of the Barents Region, the amount of nuclear pollution currently found in the region is actually fairly low (AMAP 2016, 37–39, 56; Heininen and Sergerståhl 2002; Salbu et al. 1997). The threat to the environment comes not from the current level of pollution but from the risk of future radioactive pollution caused by previous unsafe practices (Heininen and Sergerståhl 2002). Radioactive pollution in the Barents Sea as measured in both the water and the fish stocks is actually lower than in other seas but there is unsecured radioactive material in the water (Hoel 1994). The Russian Yablokov Report reported that between 1965 and 1993 seventeen nuclear reactors were placed in the Kara Sea (the most easterly sea in the Barents Region) by the Russians. In at least seven instances, the radioactive fuel was not removed from the reactor before it was dumped causing a risk that it could start to leak into the water in the future (Sivintsev et al. 2000; Yablokov et al. 1993). Likewise, large amounts of low and medium level radioactive waste were put in the sea protected only by 'flimsy metal containers' which are now in danger of corroding (Yablokov et al. 1993; Stokke 1994; Sivintsev et al. 2000). These containers are also at risk from oil and gas exploration as they are lying on the sea floor, unmarked and uncharted, and they could easily be pierced by exploratory drilling (Kireeva and Digges 2014). Lack of funding means that nuclear reactors and nuclear submarines which are no longer needed cannot be adequately decommissioned and so they, too, are at risk of deteriorating, causing nuclear material to leak (Heininen and Sergerståhl 2002). While most of the nuclear material either stored in western Russia or dumped in the seas off the Russian coast is currently contained, that containment is not secure and should nuclear material start to leak into the water or on to the land, it would pose a huge threat to the environment.

3.3.2 Maintenance of nuclear facilities

During the Cold War, there was little money for or interest in nuclear safety standards in the USSR meaning that many of the nuclear installations were of unsatisfactory quality when there were first built and have been poorly maintained ever since (Stokke 1994). A culture unconcerned with, or unable to afford, safety standards has resulted in a catalogue of recorded safety failures and accidents which have led both to fatalities and nuclear emissions which have caused, or had the potential to cause, damage to the environment (Stokke 1994). In 1989, 42 lives were lost and a nuclear reactor and two nuclear warheads went down when the Komsomólets submarine caught fire and sank near Bear Island in the Svalbard Archipelago in the Barents Sea.

The nuclear reactor has never been recovered. Regular monitoring shows that although there is currently no radiation leaking, if the casings of the reactor or the weapons were to deteriorate, the surrounding water would become contaminated (AMAP 2016, 7–8; Høibråten et al. 2003; Høibråten et al. 1997). In 2003, Russian submarine K-159 was being towed to Polyarny in order to be dismantled when it sank in the Barents Sea, taking with it all of its nuclear fuel as well as the lives of nine men whose job it was to keep water out of the already rusting hull during transportation (AMAP 2016, 33–34; Ponomareva 2008). The submarine was not seaworthy and yet it was allowed to go to sea with nuclear fuel on board. In April 2015, another nuclear submarine, the Orel, was being repaired in Arkhangelsk when it caught fire following faulty welding activity. This was the fourth such fire since 2011 to occur at shipyards owned by the United Shipbuilding company, raising questions about safety at the company's facilities (Digges 2015). While efforts are being made to reduce the threat of nuclear accidents, there remains much to do. There are, for example, currently 22,000 'spent nuclear fuel elements' being stored in 'dilapidated concrete tanks' at the Andreyeva Bay nuclear storage facility, situated only 55km from the Norwegian border (Digges 2016; Nilsen 2016b). The fuel elements were stored under water until an accident occurred in 1982 and the pool began to leak. The fuel was moved to supposedly temporary containers but, despite their poor condition, the fuel has remained in the tanks for over thirty years (Digges 2016; Nilsen 2016b). Work began in 2017 to remove the fuel elements but the task has been described as 'the most risky nuclear-safety operation ever to happen in the Russian north' and is likely to take at least five years (Nilsen 2016b). Meanwhile, and despite the internationally funded effort to clear up the damage caused by nuclear activity in the Barents Region, at least ten new nuclear submarines are being built in Russia and are due to enter service by 2020 (Digges 2014a).

3.3.3 Nuclear testing

Another source of environmental threat from nuclear activity comes from nuclear testing. The Soviet Union's nuclear testing began in 1949 and the first nuclear weapons were tested in the Barents Region in September 1955 when an underwater detonation was made close to the islands of Novaya Zemlya, which lie off the north coast of Russia, between the Barents and the Kara seas (Balonov 2004). In total, 130 nuclear tests were carried out at Novaya Zemlya between 1955 and 1990, with tests taking place in the air, on land, underground and under water (Balonov 2004; Khalturin et al. 2005; Stokke 1994; Yemelyanenkov and Popov 1992). The largest ever hydrogen bomb, a 57 megaton explosion, was tested at Novaya Zemlya in 1961 with dramatic environmental effects (Khalturin et al. 2005, 18). An observer said afterwards that the 'surface of the island has been levelled, swept and licked so that it looks like a skating rink' (Adamsky and Smirnov 1994, 3, 19–21; Suvorov 1989, 117–127).

The sheer number and size of the tests which took place at Novaya Zemlya in the second half of the twentieth century and the quantity of nuclear material released make this the largest contributor to nuclear contamination in the Arctic (United States Congress, Office of Technology Assessment 1995, 34). There is, however, some good news: in 2015 the Arctic Monitoring and Assessment Programme found that 'levels of anthropogenic radionuclides' in the Barents Sea were lower than had been recorded between the 1970s and the 1990s and that levels were continuing to fall (AMAP 2016, 56). As long as no new nuclear tests are carried out, the levels of radionuclides will naturally continue to fall over time.

3.3.4 Nuclear reactor construction in Finland

In northern Finland, there is a controversial plan in place to build a nuclear power plant in the Pyhäjoki municipality (World Nuclear Association 2017). Preliminary construction work on the Hanhikivi plant has begun and an application for a construction permit, requesting permission to construct a nuclear power plant, nuclear fuel storage facilities, nuclear waste processing facilities and interim storage facilities for spent nuclear fuel, was submitted by the company, Fennovoima, in 2015 (Fennovoima 2015). The company had hoped to begin construction in 2018 with the power plant due to be operational by 2024 but the Finnish nuclear safety agency has reported delays to its decision regarding the safety of the plans (Yle 2017; Fennovoima 2015, 10–11).

In 2015, an Environmental Impact Assessment undertaken by Fennovoima claimed that 'the project will have no adverse environmental impacts that would be unacceptable or could not be mitigated to an acceptable level' (Fennovoima 2015, 10). Despite this, the plan has been criticised by environmental groups who have concerns about the impact it will have on the environment (Digges 2014b; Goës 2015). The plant will be located 2km from the Parhalahti–Syolatinlahti ja Heinikarinlampi Natura 2000 area which is designated to protect 29 species and 8 habitats (European Environment Agency 2017; Fennovoima 2015, 64; Natura 2000 2017). The site is also in the vicinity of a Finnish Important Bird Area and is home to a large quantity of nesting wetlands species. In addition, there are also a number of nature conservation areas, particularly seashore meadows, nearby.

The building of any industrial plant in the area would be likely to lead to biodiversity loss and habitat degradation; a nuclear power plant with the added risks of exposure to radiation could be disastrous for the environment (Fennovoima 2015, 74–75; Kahsnitz 2016, 22). More concerning is the predicted impact on the seawater close to the plant. The plant operators intend to use seawater as a coolant and to discharge the used water back into the Bay of Bothnia. It is predicted that this discharge will raise the sea temperature by up to 10–12°C in the immediate area and up to 5°C in an area up to 0.7km^2 around the discharge site (Fennovoima 2015, 68–71). The warm water will shorten the period of time for which the sea is frozen during the winter and will result in thinner and weaker sea ice forming (Fennovoima 2015, 69). The temperature rise is likely to lead to changes in species distribution, changes to marine vegetation and bring an end to ice fishing (Fennovoima 2015, 68–71). There is a risk that the seal population could be threatened as the reduction in sea ice levels will affect their ability to breed and, as the primary predator in the region, seals will be at the greatest risk of bioaccumulation (Kahsnitz 2016, 23). Despite Fennovoima's assurances that the plant will have only a limited environmental impact, it is clear that there are considerable environmental concerns associated with the construction of the nuclear power plant and that the plant could pose a threat to the environmental security of the area and the broader region (Fennovoima 2015, 10, 62, 85).

3.4 Environmental hot spots

Recognising the threats posed to the environment in the Barents Region, the Barents Euro-Arctic Council's Working Group on the Environment identified a number of sites which were of greatest concern when considering environmental threats in the region (AMAP 1995). They called these areas 'hot spots' and began work to reduce the impact of the pollution caused by these sites (AMAP 1995). The work was reviewed in 2003 when a new list of 42 'hot spots' was drawn up (AMAP 2003). The hot spots represent a wide range of different environmental and health threats including industrial air emissions, discharges of waste water, threats to drinking water supplies, waste management and other environmental damage (Bambukyak et al. 2013). Work undertaken to ameliorate the hot spots includes the removal of twenty sunken and abandoned ships near Lavina in the Kola Bay and the beginning of work to remove 18,000 tonnes of metal from the 181 islands of the Franz Joseph Land archipelago (Mård 2013, 34–37, 42–45). Since 2011, nine of the hot spots have been excluded from the list because they are considered to have improved to such an extent that they no longer pose a threat to the environment (Bambukyak et al. 2013, 13, 18; Barents Euro-Arctic Council 2011, 2013c, 2015). One of the 'hot spots' which has been excluded is the stock of obsolete pesticides in the Republic of Karelia which, as a result of successful cooperation, were incinerated in Finland as the technology to render them harmless was not available in Russia (Bambukyak et al. 2013, 18, 55–56).

The work on the identified 'hot spots' has led to significant improvements in the air quality or water quality and recovery of the environment in the areas around the sites. There is, however, much work still to do: in the fourteen years since the 42 'hot spots' were first identified it has only been possible to exclude nine of the sites from the list, with 33 sites still causing concern (Bambukyak et al. 2013; Barents Euro-Arctic Council 2011, 2013c; Mikaelsson 2015). Additional work is still required on these remaining sites. Furthermore, the list only included the sites which were the most concerning in 2003; there are still many more sites within the region which are actively polluting or threatening to pollute the environment and work must be undertaken to ameliorate these sites as well as those which made the list of 'hot spots'.

4 Conclusion and recommendations

Having discussed the various threats to the environment of the Barents Region, it is worth considering some of the possible solutions to those threats. These proposals do not seek to provide all of the answers to the environmental threats identified but instead give some ideas about action that could be taken in light of the issues highlighted by this chapter.

Many of the worst sites of pollution within the Barents Region have been transformed by the work undertaken after their designation as environmental 'hot spots'. There are, however, many more sites which did not make it on to the list but which remain an environmental threat. It is important that once the worst polluters have been dealt with, focus should turn to other polluting sites. One solution to this would be to expand the environmental 'hot spots' programme, either through an increase in the number of sites on the list or by selecting a new site to replace each site which is eliminated from the list. This would ensure that work to tackle industrial pollution and nuclear clean up in the Barents Region will continue. The benefit of a 'one out, one in' system is that it would keep the level of work required to manage the programme and the overall annual costs, currently financed through the Nordic Environmental Financing Corporation, consistent, while continuing and extending the valuable work undertaken by the project to date.

Despite the claims of its owners, the proposed nuclear power plant in Pyhäjoki, Finland, is likely to have a considerable impact on the environment, particularly on the marine environment in the Bay of Bothnia. While Finland's commitment to energy sources which do not cause greenhouse gas emissions is laudable, Finland should be investing in renewable energy sources rather than building nuclear power plants. Finland has the natural resources to enable it to move towards renewable energy and it should be directing its time, finances and expertise towards energy sources which are less environmentally harmful and do not put future generations at risk of environmental insecurity.

There is, as always, an urgent need for action on climate change. Despite acknowledging the threat posed by climate change, the Barents Euro–Arctic Council has done little to take action on the subject. In December 2013 an 'Action Plan on Climate Change' was adopted (Barents Euro-Arctic Council 2013a) but the plan is surprisingly weak, partly because of limited resources within the Barents cooperation and partly because the council reached the conclusion that '[t]he Barents Region is not a major source of greenhouse gas emissions' (Barents Euro-Arctic Council 2013b). The action plan does little more than recommend that each part of the Barents Region produces its own climate strategy. Lack of resources is not a good enough reason for the Barents Region to fail to produce a detailed plan or even a binding agreement between the Member States on reducing greenhouse gas emissions. A strong, region-wide strategy on climate change is required for three reasons: to encourage the reduction of greenhouse gas emissions, to make plans for adapting to the consequences of climate change in the Arctic, and in order provide an example for other countries. Climate change ought to be a priority for all Arctic nations and by not having in place a robust climate plan, the Barents Region appears to be failing to take the threat sufficiently seriously.

If the countries of the Barents Region were to focus on reducing their emissions of black carbon, this would have a direct effect on climate change in the region. The warming effect of black carbon emissions is localised so action taken in the Barents Region to reduce emissions would have a direct effect in reducing local temperature rises. At the very least, the practice of gas flaring should cease in favour of a process known as 'recovery and utilisation' where the gases are collected and reused rather than being flared (AMAP 2015, 33). The problem with this is that the build-up of waste hydrocarbon gas can be dangerous and flaring is often used as a safety mechanism (AMAP 2015, 33). A solution has been found in Norway where gas flaring is banned unless it is specifically required for security reasons (Arctic Council 2013, 8; Norwegian Act Relating to Petroleum Activities 1996, sec. 4.4). Because of the effect that black carbon has on increasing the temperature in the Arctic, this is one area where the countries in the Barents Region can affect the rate of warming in the Arctic rather than requiring global action. As such, measures should be taken across the region to reduce and then ban gas flaring except in instances when it is required for safety.

This chapter defined environmental security as 'the protection of the natural environment to ensure the sustainable functioning of ecosystems and biodiversity both for the intrinsic value of nature and for the support of the survival, health and livelihoods of both individuals and communities'. The environmental issues which have been identified by this chapter threaten the environmental security of the Barents Region by damaging habitats, by reducing biodiversity, by destroying or reducing access to food, clean water and clean air and by impairing local people's ability to make a living from the environment. When the environment becomes polluted or damaged, food sources will become contaminated or will disappear, livelihoods which rely on the environment, ranging from reindeer herding to tourism, will be destroyed, local people will suffer from sickness and disease and traditional ways of life will be threatened. This creates communities which are not environmentally secure and which are therefore vulnerable. It is vital that we protect the environment of the Barents Region, partly because of the inherent value of nature and the environment but also because without environmental security, there can be no human security.

References

Adamsky, V. and SmirnovY. 1994. Moscow's biggest bomb: the 50-megaton test of October 1961. *Cold War International History Project Bulletin*Issue 4. Washington DC, USA: Woodrow Wilson International Center for Scholars.

AMAP. 1995. *Report to the Nordic Environment Finance Corporation (NEFCO). Proposals for Environmentally Sound Investment Projects in the Russian Part of the Barents Region*. Oslo, Norway: Arctic Monitoring and Assessment Programme. www.amap.no/documents/doc/amap-report-to-nefco.-proposals-for-environmentally-sound-investment-projects-in-the-russian-part-of-the-barents-region/837.

AMAP. 2003. *Updating of Environmental Hot Spots List in the Russian Part of the Barents Region: Proposal for Environmentally Sound Investment Projects*. Oslo, Norway: Arctic Monitoring and Assessment Programme. www.amap.no/documents/doc/updating-of-environmental-hot-spots-list-in-the-russian-part-of-the-barents-region/838.

AMAP. 2005. *Arctic Climate Impact Assessment. ACIA Overview Report*. Cambridge, UK and New York, USA: Cambridge University Press. www.amap.no/documents/doc/arctic-arctic-climate-impact-assessment/796.

AMAP. 2006. *Arctic Pollution 2006: Acidification and Arctic Haze*. Oslo, Norway: Arctic Monitoring and Assessment Programme. www.amap.no/documents/doc/amap-assessment-2006-acidifying-pollutants-arctic-haze-and-acidification-in-the-arctic/91.

AMAP. 2011. *Snow, Water, Ice and Permafrost in the Arctic (SWIPA): Climate Change and the Cryosphere*. Oslo, Norway: Arctic Monitoring and Assessment Programme. www.amap.no/documents/doc/arctic-climate-issues-2011-changes-in-arctic-snow-water-ice-and-permafrost/129.

AMAP. 2015. *AMAP Assessment 2015: Black Carbon and Ozone as Arctic Climate Forcers*. Oslo, Norway: Arctic Monitoring and Assessment Programme. www.amap.no/documents/doc/amap-assessment-2015-black-carbon-and-ozone-as-arctic-climate-forcers/1299.

AMAP. 2016. *AMAP Assessment 2015: Radioactivity in the Arctic*. Oslo, Norway: Arctic Monitoring and Assessment Programme. www.amap.no/documents/doc/AMAP-Assessment-2015-Radioactivity-in-the-Arctic/1457.

AMAP. 2017a. *Adaptation Actions for a Changing Arctic: Barents Area Overview Report*. Oslo, Norway: Arctic Monitoring and Assessment Programme. www.amap.no/documents/doc/Adaptation-Actions-for-a-Changing-Arctic-AACA-Barents-Area-Overview-report/1529.

AMAP. 2017b. *Snow, Water, Ice and Permafrost in the Arctic (SWIPA): Summary for Policymakers*. Oslo, Norway: Arctic Monitoring and Assessment Programme. www.amap.no/documents/doc/Snow-Water-Ice-and-Permafrost.-Summary-for-Policy-makers/1532.

Arctic Council. 2013. *Recommendations to Reduce Black Carbon and Methane Emissions to Slow Arctic Climate Change*. Kiruna, Sweden: Arctic Council. https://oaarchive.arctic-council.org/handle/11374/80.

Balonov, M. 2004. *Nuclear Explosions in the USSR: The North Test Site, Reference Material*. Version 4. Vienna, Austria: International Atomic Energy Agency.

Bambukyak, A., Golubeva, S. and Savinov, V. 2013. *Assessment of the Barents Hot Spot Report Describing the State of 42 Original Barents Environmental Hot Spots. Part I – Analysis*. Nordic Environment Finance Corporation and Barents Hot Spot Facility.

Barents Euro-Arctic Council. 1994. *First Meeting of the Ministers of the Environment: Barents Region Environment Action Programme*. Bodø, Norway: Barents Euro-Arctic Council.

Barents Euro-Arctic Council. 2011. *Tenth Meeting of the Ministers of the Environment: Declaration*. Umeå, Sweden: Barents Euro-Arctic Council.

Barents Euro-Arctic Council. 2013a. *Barents Euro-Arctic Council: Action Plan on Climate Change for the Barents Cooperation*. Inari, Finland: Barents Euro-Arctic Council.

Barents Euro-Arctic Council. 2013b. *Declaration on the 20th Anniversary of the Barents Euro-Arctic Cooperation*. Kirkenes, Norway: Barents Euro-Arctic Council.

Barents Euro-Arctic Council. 2013c. *Eleventh Meeting of the Ministers of the Environment: Declaration*. Inari, Finland: Barents Euro-Arctic Council.

Barents Euro-Arctic Council. 2015. *Twelfth Meeting of the Ministers of the Environment: Declaration*. Sortavala, Russian Federation: Barents Euro-Arctic Council.

Barents Euro-Arctic Council. 2017. Population and culture. *Barents Cooperation*. www.barentscooperation.org/en/About/Barents-region/population.

Barents Info. 2017. Maps from the Barents Region. *Barents Info Portal*. www.barentsinfo.org/Contents/Maps.

Bedford, T. 2015. Talvivaara Mine Environmental Disaster. *Environmental Justice Organisations, Liabilities and Trade Factsheet*, no. 37.

Butler, C. 2016. *Climate Change and Global Health*. Wallingford, UK and Boston, MA, USA: CABI.

Collins, M., Knutti, R., Arblaster, J.*et al.*2013. Long-term climate change: projections, commitments and irreversibility. In *Climate Change 2013: The Physical Science Basis. Contribution of Working Group I to the Fifth Assessment Report of the Intergovernmental Panel on Climate Change*, edited by T.F. Stockeret al. Cambridge, UK and New York, USA: Cambridge University Press.

Coskun, H. Gonca, Kerem Cigizoglu, H. and Derya Maktav, M. 2008. *Integration of Information for Environmental Security*. Dortrecht, The Netherlands: Springer.

Digges, C. 2014a. Soaring Russian nuclear submarine construction dwarfs post-soviet dismantlement spending. *Bellona.org*, July 28. http://bellona.org/news/nuclear-issues/radioactive-waste-and-spent-nuclear-fuel/2014-07-soaring-russian-nuclear-submarine-construction-dwarfed-post-soviet-dismantlement-spending.

Digges, C. 2014b. Finnish demonstrators gather to block construction of Russian-funded nuclear power plant. *Bellona.org*, October 17. http://bellona.org/news/nuclear-issues/nuclear-russia/2014-10-finnish-demonstrators-gather-block-construction-russian-funded-nuclear-power-plant.

Digges, C. 2015. Russian nuclear submarine still smoldering after serious fire in dry-dock. *Bellona.org*, April 7. http://bellona.org/news/nuclear-issues/2015-04-russian-nuclear-submarine-still-smoldering-serious-fire-dry-dock.

Digges, C. 2016. Andreyeva Bay – a history of successful international cooperation. *Bellona.org*, October 26. http://bellona.org/news/nuclear-issues/radioactive-waste-and-spent-nuclear-fuel/2016-10-andreyeva-bay-a-history-of-successful-international-cooperation.

European Environment Agency. 2017. European nature information system: site fact-sheet for Parhalahti-Syölätinlahti Ja Heinikarinlampi. *European Environment Agency*. http://eunis.eea.europa.eu/sites/FI1104201.

Fennovoima. 2015. *Application for a Construction License pursuant to Section 18 of the Nuclear Energy Act (990/1987) for the Hanhikivi 1 Nuclear Power Plant*. https://issuu.com/fennovoima/docs/construction_license_application_pu.

Förstner, U., and Wittmann, G.T.W. 2012. *Metal Pollution in the Aquatic Environment*. Berlin, Germany, Heidelberg, Germany, New York, USA: Springer.

Fryer, P., Brown-Leonardi, C. and Soppela, P. 2010. *Encountering the Changing Barents*. Rovaniemi, Finland: Arctic Centre, University of Lapland.

Glenn, J.C., Gordon, T.J. and Perelet, R. 1997. Environmental security: emerging international definitions, perceptions, and policy considerations. In *2001 State of the Future*, edited by J.C. Glenn and T.J. Gordon. The Millennium Project.

Goës, E. 2015. Swedish protests against the plans for a nuclear plant in Pyhäjoki (Finland). *Global Greens*, October 22. www.globalgreens.org/news/swedish-protests-against-plans-nuclear-plant-pyh äjoki-finland.

Granqvist, K. 2014. The mining situation in Sweden from an environmental perspective – a few examples. *Arctic Anthropology* (blog), January 13. https://arcticanthropology.org/2014/01/13/the-mining-situation-in-sweden-from-an-environmental-perspective-a-few-examples/.

Granqvist, K. 2016. Historical decision as to mining in Sweden. *Arctic Anthropology* (blog), May 2. https://arcticanthropology.org/2016/05/02/historical-decision-as-to-mining-in-sweden/.

Heininen, L. and Sergerståhl, B. 2002. International negotiations aiming at reduction of nuclear risks in the Barents Region. In *Containing the Atom*, edited by R. Avenhaus, V. Aleksandrovich Kremeniuk, and G. Sjöstedt. Lanham, MD, USA: Lexington Books.

Hoel, A. 1994. The Barents Sea. In *The Barents Region: Cooperation in Arctic Europe*, edited by O. Schram Stokke and O. Tunander. London, UK: Sage Publications.

Høibråten, S., Haugan, A. and Thoresen, P. 2003. *The Environmental Impact of the Sunken Submarine Komsomolets*. Norway: Forsvarets Forskningsinstitutt, Norwegian Defence Research Establishment.

Høibråten, S., ThoresenP., and Haugan, A. 1997. The Sunken nuclear submarine Komsomolets and its effects on the environment. *The Science of the Total Environment* 202(1–3): 67–78.

Hull, R.N., Barbu, C.H. and Goncharova, N. 2009. *Strategies to Enhance Environmental Security in Transition Countries*. Dordrecht, The Netherlands: Springer.

IPCC. 2007. *IPCC, 2007: Climate Change 2007: Synthesis Report. Contribution of Working Groups I, II and III to the Fourth Assessment Report of the Intergovernmental Panel on Climate Change: Summary for Policymakers*, edited by R.K. Pachauri, A. Reisinger and Core Writing Team. Geneva, Switzerland: Intergovernmental Panel on Climate Change.

Kahsnitz, J. 2016. *The Construction of the Nuclear Power Plant Hanhikivi 1 – Implications for Bothnian Bay Fauna*. Gothenburg, Sweden: University of Gothenburg.

Kashulina, G., Reimann, C., Finne, T. E., Halleraker, J.H., Äyräs, M. and Chekushin, V.A. 1997. The state of the ecosystems in the Central Barents Region: scale, factors and mechanism of disturbance. *Science of The Total Environment* 206(2): 203–225.

Khalturin, V. I., Rautian, T.G., Richards, P.G. and Leith, W.S. 2005. A review of nuclear testing by the Soviet Union at Novaya Zemlya, 1955–1990. *Science & Global Security* 13(1–2): 1–42.

Kireeva, A. and Digges, C. 2014. Wanted: Someone to decide what to do with Russia's radioactive debris at the bottom of the Arctic. *Bellona.org*, September 23. http://bellona.org/news/nuclear-issues/radioactive-waste-and-spent-nuclear-fuel/2014-09-wanted-someone-decide-russias-radioactive-debris-bottom-arctic.

Klima-og miljødepartementet. 2016. *Opprettholder tillatelse til gruvedrift i Nussir og Ulveryggen*. https://www.regjeringen.no/no/aktuelt/opprettholder-tillatelse-til-gruvedrift-i-nussir-og-ulveryggen/id2524800/

Kozlov, M. V. and Barcan, V. 2000. Environmental contamination in the central part of the Kola Peninsula: history, documentation, and perception. *AMBIO: A Journal of the Human Environment* 29(8): 512–517.
Kryuchkov, V. 1990. Extreme anthropogenic load and the state of the North Taiga ecosystem. In *Effects of Air Pollutants and Acidification in Combination with Climatic Factors on Forests, Soils, and Waters in Northern Fennoscandia*, edited by K. Kinnunen and M. Varmola. Copenhagen, Denmark: Nordic Council of Ministers.
Labba, K. 2015. The legal organisation of Sami reindeer herding. In *Indigenous Rights in Scandinavia: Autonomous Sami Law*, edited by C. Allard and S. Funderud Skovang. Farnham, UK: Ashgate Publishing.
Larson, J.N., Anisimov, O.A., Constable, A., Hallowed, A.B., Maynard, N., Prestrud, P., Prowse, T.D. and Stone, J.M.R. 2014. Polar regions. In *Climate Change 2014: Impacts, Adaptation, and Vulnerability. Part B: Regional Aspects. Contribution of Working Group II to the Fifth Assessment Report of the Intergovernmental Panel on Climate Change*, edited by V.R. Barros*et al.*Cambridge, UK and New York, USA: Cambridge University Press, 1567–1612.
Lemly, A.D. 1996. Wastewater discharges may be most hazardous to fish during winter. *Environmental Pollution* 93(2): 169–174.
Malinovsky, D., Rodushkin, I., Moiseenko, T. and Öhlander, B. 2002. Aqueous transport and fate of pollutants in mining area: a case study of Khibiny Apatite–nepheline Mines, the Kola Peninsula, Russia. *Environmental Geology* 43(1–2): 172–187.
Mård, A. 2013. Hot spots: tacking environmental challenges in the Barents Region. Edited by M. Sjövall. Tromsø, Norway: Nordic Environment Finance Corporation and Barents Hot Spot Facility. www.nefco.org/sites/nefco.org/files/pdf-files/barents_hot_spots_final.pdf.
Mikaelsson, Å. 2015. *Progress Achieved During Implementation of the Hot Spots Exclusion Procedure. Presentation at the Twelfth Meeting of the Ministers of the Environment, Barents Euro-Arctic Council, Sortavala, Russian Federation.* Presented at the Twelfth Meeting of the Ministers of the Environment, Barents Euro-Arctic Council, Sortavala, Russian Federation, November 25.
Mines and Communities. 2016. Finland: Talvivaara bosses merely fined for 'environmental crimes'. *Mines and Communities*, May 13. www.minesandcommunities.org/article.php?a=13385.
Moiseenko, T. I. and Kudryavtseva, L.P. 2001. Trace metal accumulation and fish pathologies in areas affected by mining and metallurgical enterprises in the Kola Region, Russia. *Environmental Pollution* 114(2): 285–297.
Moiseenko, T. I. 1999. The fate of metals in Arctic surface waters. Method for defining critical levels. *Science of The Total Environment* 236(1–3): 19–39.
National Snow and Ice Data Center. 2016. Sea ice hits record lows. National Snow and Ice Data Center, December 6. http://nsidc.org/news/newsroom/sea-ice-hits-record-lows.
Natur og Ungdom and Naturvernforbundet. 2015. *Complaint for Permission for Activities under the Pollution Act – Nussir ASA: Natur Og Ungdom and Naturvernforbundet.* www.regjeringen.no/contentassets/0da96c16ad2b49f59afc736215ac1056/klage-fra-naturvernforbundet-natur-og-ungdom-l321191.pdf.
Natura 2000. 2017. Natura 2000 Designation Parhalahti-Syölätinlahti Ja Heinikarinlampi. Europa, February 2. http://natura2000.*eea*.europa.eu/Natura2000/SDF.aspx?site=FI1104201.

Nieminen, P., Panychev, D., Lyalyushkin, S.et al. 2013. Environmental exposure as an independent risk factor of chronic bronchitis in northwest Russia. *International Journal of Circumpolar Health* 72(1): 19742.

Nilsen, T. 2014. Very strange to us that you will dump mining waste at sea. *The Independent Barents Observer*, May 20. http://barentsobserver.com/en/nature/2014/05/very-strange-us-you-will-dump-mining-waste-sea-20-05.

Nilsen, T. 2016a. Norway votes against ban on dumping mining waste at sea. *Independent Barents Observer*, October 22. https://thebarentsobserver.com/en/ecology/2016/10/norway-votes-against-ban-dumping-mining-waste-sea.

Nilsen, T. 2016b. Arctic's most dangerous Cold War legacy soon to be removed. *Independent Barents Observer*, October 26. https://thebarentsobserver.com/en/ecology/2016/10/arctics-most-dangerous-cold-war-legacy-soon-be-removed.

Norseth, T. 1994. Environmental pollution around nickel smelters in the Kola Peninsula (Russia). *Science of The Total Environment, Nickel Biochemistry, Toxicology, and Ecologic Issues* 148(2): 103–108.

Norwegian Act Relating to Petroleum Activities. 1996.

Nuttall, M. 2007. An environment at risk: Arctic indigenous peoples, local livelihoods and climate change. In *Arctic Alpine Ecosystems and People in a Changing Environment*, edited by J. Børre Ørbæk*et al.*Berlin, Heidelberg, New York: Springer.

Olson, D.M., Dinerstein, E., Wikramanayake, E.D.*et al.*2001. Terrestrial ecoregions of the world: a new map of life on earth. *BioScience* 51(11): 933–938.

Pachauri, R.K., Meyer, L.A. and Core Writing Team (Eds). 2014. *IPCC, 2014: Climate Change 2014: Synthesis Report. Contribution of Working Groups I, II and III to the Fifth Assessment Report of the Intergovernmental Panel on Climate Change.* Geneva, Switzerland: Intergovernmental Panel on Climate Change. www.ipcc.ch/report/ar5/syr/.

Pasanen, K., Pukkala, E., Turunen, A.W.*et al.*2012. Mortality among population with exposure to industrial air pollution containing nickel and other toxic metals. *Journal of Occupational and Environmental Medicine* 54(5): 583–591. doi:10.1097/JOM.0b013e3182492050.

Plummer, R. and BairdJ. 2013. Adaptive co-management for climate change adaptation: considerations for the Barents Region. *Sustainability* 5(2): 629–642.

Ponomareva, V. 2008. Five years of promises to raise the K-159 remain broken. Translated by C. Digges. *Bellona.org*, September 1. http://bellona.org/news/nuclear-issues/accidents-and-incidents/2008-09-five-years-of-promises-to-raise-the-k-159-remain-broken.

Przybylak, R. 2015. *The Climate of the Arctic*. 2nd ed. Switzerland: Springer International Publishing.

Pulkkinen, E. (Ed.). 1991. Heavy metal accumulation in spruce needles and changes of Northern Taiga ecosystems. *Environmental Geochemistry in North Europe. Geological Survey of Finland*, Special Paper 9, 177–184.

Rekacewicz, P. 2005a. Air temperature in the Barents Region in winter. *Barentswatch Atlas.* www.grida.no/graphicslib/detail/air-temperature-in-the-barents-region-in-winter_9ce2.

Rekacewicz, P. 2005b. Forest damage due to air pollution. *Barentswatch Atlas.* www.grida.no/graphicslib/detail/forest-damage-due-to-air-pollution_80fc.

Revich, B.A. 1995. Public health and ambient air pollution in Arctic and Subarctic cities of Russia. *Science of The Total Environment* 160–161: 585–592.

Russian Security Council. 1996. *Environmental Security of Russia* (issue 2). Moscow: Russian Federation.

Salbu, B., Nikitin, A. I., Strand, P.et al. 1997. Radioactive contamination from dumped nuclear waste in the Kara Sea. *Science of The Total Environment* 202(1): 185–198.

Sand, M., Koren Berntsen, T., Seland, Ø. and Kristjánsson, J.E. 2013. Arctic surface temperature change to emissions of black carbon within Arctic or midlatitudes. *Journal of Geophysical Research: Atmospheres* 118(14): 7788–7798.

Siberian Times. 2016. Child, 12, died from anthrax, as nine cases confirmed of the deadly disease. August 1. http://siberiantimes.com/other/others/news/n0693-eight-people-have-contracted-anthrax-amid-reports-that-one-has-died/.

Sivintsev, Yu.V., Vakulovsky, S.M., Vasiliev, A.P.*et al.*2000. *Technogenic Radionuclides in the Sea Surrounding Russia. Radioecological Consequences of Radioactive Waste Dumping in the Arctic and Far Eastern Seas. The White Book – 2000*. Moscow: Russian Federation.

Solomon, S., Qin, D., ManningM.*et al.* (Eds). 2007. *Climate Change 2007: The Physical Science Basis, Contribution of Working Group I to the Fourth Assessment Report of the Intergovernmental Panel on Climate Change*. Cambridge, UK and New York, USA: Cambridge University Press.

Staalesen, A. 2016. Government gives thumbs up for mining company, will be allowed to dump wastes in Arctic fjord. *Independent Barents Observer*, December 20. https://thebarentsobserver.com/en/ecology-industry-and-energy/2016/12/government-gives-thumbs-mining-company-will-be-allowed-dump.

Stammler, F. 2005. *Reindeer Nomads Meet the Market: Culture, Property and Globalisation at the End of the Land*. Münster, Germany: LIT Verlag Münster.

Stohl, A., Klimont, Z., Eckhardt, S.*et al.*2013. Black carbon in the Arctic: the underestimated role of gas flaring and residential combustion emissions. *Atmospheric Chemistry and Physics* 13(17): 8833–8855.

Stokke, O.S. 1994. Environmental cooperation as a driving force in the Barents Region. In *The Barents Region: Cooperation in Arctic Europe*, edited by O.S. Stokke and O. Tunander. London, UK: Sage Publications.

Stonehouse, B. and Snyder, J. 2010. *Polar Tourism: An Environmental Perspective*. Bristol, UK: Channel View Publications.

Suvorov, V.A. 1989. *Strana Limoniia (Land of Lemons)*. Moscow, Russian Federation: Sovetskaia Rossiia.

Tuovinen, J.-P., Laurila, T., Lättilä, H., Ryaboshapko, A., Brukhanov, P. and Korolev, S. 1993. Impact of the sulphur dioxide sources in the Kola Peninsula on air quality in northernmost Europe. *Atmospheric Environment* 27(9): 1379–1395.

UN Development Programme. 1994. *Human Development Report*. Oxford: Oxford University Press.

UNESCO. 2006. *State of Conservation of World Heritage Properties in Europe: Sweden Laponian Area*. UNESCO. http://whc.unesco.org/archive/periodicreporting/EUR/cycle01/section2/774-summary.pdf.

US Office of Technology Assessment. 1995. *Nuclear Wastes in the Arctic*. Washington, DC: US Government Printing Office.

World Nuclear Association. 2017. Nuclear energy in Finland. World Nuclear Association, January 31. www.world-nuclear.org/information-library/country-profiles/countries-a-f/finland.aspx.

Yablokov, A., Karasev, V., Rumyantsev, V.*et al.*1993. *Yablokov Commission Report. Facts and Problems Related to Radioactive Waste Disposal in Seas Adjacent to the Territory of the Russian Federation*. Report to the President of the Russian Federation. Moscow, Russian Federation.

Yemelyanenkov, A. and Popov, V. 1992. *Atom Bez Grifa sekretno (Atoms without the Stamp of Secrecy: Viewpoints)*. Moscow, Russian Federation and Berlin, Germany: International Physicians for the Prevention of Nuclear War.

Yle. 2017. Fennovoima nuclear project faces delays over safety clearance, building permit. *Yle Uutiset*, May 19. https://yle.fi/uutiset/osasto/news/daily_fennovoima_nuclear_project_faces_delays_over_safety_clearance_building_permit/9622096.

2.2 Economic security in the Barents Region

Anna Petrétei and Dorothée Cambou

This chapter analyses economic security in the Barents Region. Economic security is one of the indicators for human security (UNDP 1994, 24), and is intrinsically linked with other human security components.

1 Definition

Economic security is composed of basic social security, and defined by access to basic needs infrastructure pertaining to health, education, dwelling, information, social protection, and work-related security (ILO 2004). According to the UN Human Development Report, economic security requires an assured basic income for individuals, usually from productive and remunerative work or, as a last resort, from a publicly financed safety net. In this sense, the majority of the world's population may be deemed as economically insecure (UNDP 1994, 25). This is mostly due to the unstable world market, increasing job insecurity, and growing unemployment ratio. The situation is even worse in developing countries, where governments have less means to provide economic support. The UNDP also identifies disparities in economic opportunities as one of the emerging threats that constitutes the global framework of human insecurity (UNDP 1994, 34–35). Thus, economic security is closely intertwined with job and income security. Income security denotes adequate actual, perceived, and expected income, either earned or in the form of social security and other benefits (ILO 2004). For some communities such as local and indigenous peoples informal income also represents a substantial share of the income.

Several other factors contributing to income and economic security include: a satisfying employment rate, an adequate government financed social safety net, and a stable market. Employment rate is defined by the OECD as a measure of the extent to which available labour resources are being used. It is the ratio of the employed to the working age population.[1] This indicator is seasonally adjusted and it is measured in terms of thousand persons aged 15 and over, and as a percentage of working age population. Employment rate is sensitive to economic cycles, and can be significantly impacted by existing government policies and public demographics. This includes the level of higher education achievable, the income support policies of governments, and policies that facilitate the employment of women and disadvantaged groups. Another contributing factor to economic security is the existence of a social safety net, defined as a collection of services provided by the state or other institutions such as friendly societies, including cash or food transfers, social services, such as health clinics; and insurance options (Gentilini & Omamo 2009, 2). Adequate government financed social safety nets often include maternity leave, welfare, unemployment benefit, universal healthcare, pension, homeless shelters, and sometimes subsidised services such as public transport, which are meant to prevent individuals from falling into poverty beyond a certain level. The UNDP suggests that a means of determining whether social safety is adequately ensured includes measuring homelessness as an indicator of insecurity (UNDP 1994, 26; UNDP 2016, 55).

When discussing economic security, the stability of the market is an important indicator to take into account. While it is difficult to appraise the stability of the market economy, several indicators may provide information about the economic situation of a country such as its GDP (Gross Domestic Product) and GRDP (Gross Regional Domestic Product). GDP embodies the market value of all products and services produced within a certain period, by all people and companies in the given country, whereas GRDP is the geographic breakdown of national GDP, indicating the size and structure of regional economies and measuring changes to regional economies over time. Although GDP is one of the most effective ways to measure the economic development of a country, GDP and GRDP are not good indicators of economic security *per se*. This is because GDP and GRDP are the main indicators of economic growth. Whereas economic security aims to ensure economic wellbeing, economic growth is a phenomenon of market productivity and rise in GDP or GRDP. Hence, as economist Amartya Sen indicates, 'economic growth is one aspect of the process of economic development', but it does not subsume such development (Sen 1983, 748).

Understanding that economic growth alone cannot properly determine the security or development of a country, it is therefore necessary to consider other indicators. A more inclusive indicator of economic development constitutes the human development indicator. The creation of the Human Development Index (HDI) emphasises that people and their capabilities should be the ultimate criteria for assessing the development of a country. The HDI is a summary measure of average achievements in three key dimensions of human development: a long and healthy life, education, and a decent standard of living, the latter being measured on the basis of the GDP per capita. The use of HDI overlaps with other aspects of human security as well, but in the case of economic security, it serves a more representative function than economic growth. However, several important aspects of human development are also not well represented in HDI, especially as it is difficult to use indicators like GDP per capita to measure the health of subsistence or mixed economies (AHDR 2004, 17). In this regard, HDI can be a valuable tool to assess economic security but certainly not the only one.

Lastly, addressing economic security requires interdisciplinary actions across sectors, communities, and political borders in order to ensure a more comprehensive overview of the matter. In the globalised free market, interaction between economies is essential and unavoidable, which consequently requires stable and peaceful political and economic relations across borders. In summary, economic security is far-reaching, and includes the need to address the relationships between various security aspects in both contemporary and future situations.

2 Contextualisation

The Barents Region has considerable economic potential and is exceptionally rich in natural resources. The economy of the Barents is also strong, with an overall positive GDP since the beginning of the 21st century (BEAC 2015, 14). Important drivers of the Barents Region's economy are forestry, metallurgy, mining, energy production, fisheries, and tourism. In terms of primary sectors, forestry and mining are important in all parts of the Barents Region, while fishing and energy (mainly oil and gas) are important in northern Norway and northwestern Russia. The Barents Region is also an important source of hydroelectricity and a growing market for wind energy, for both local use and for export outside the region. Tourism and reindeer husbandry are also important activities in the region. In some areas, such as northern Finland, the sector of tourism is an important source of employment, and its importance is growing in other parts of the region, especially in Norway (BIN 2017). In addition, the secondary workforce sector – processing, production, and construction – accounts for most work in Fennoscandia and some Russian areas. In the Nordic part of the Barents area, the services sector is the largest employer (AMAP 2017).

There are, however, still apparent economic divisions between the Nordic and Russian parts of the region as the Nordic countries enjoy substantially higher GDP per capita than Russian regions. In addition, it can be noted that there is also variation between Barents and non-Barents Regions within the same country. In the Northern areas of Norway, Sweden, and Finland, per capita economic growth has generally lagged behind the respective national averages (AMAP 2017, 11). This is mostly due to less diverse industrial bases and lower labour market participation (AMAP 2017, 11). In the Russian Federation on the other hand, the resource-rich Northern regions have the highest GDP per head compared to the national average (AMAP 2017, 11). On the other hand, Nordic countries currently enjoy substantially higher GDP per capita than regions in the Russian Federation.

Yet, the Barents Region has encountered the consequences of economic recession and looking at GRP is insufficient to assess the reality concerning the basis for economic livelihoods. In the Barents Region, job creation is generally lagging behind the national average and employment creation in the resources sector has significantly decreased since 2008 (BIN 2017). In the Nordic countries, the biggest job loss between 2008 and 2014 was observed in mining, quarrying and manufacturing (4,444 jobs), followed by agriculture, forestry, and fishing (2,270 jobs) (BIN 2017, 45). According to Business Index North (BIN), 'job losses in agriculture, forestry and fishing reflect the process of urbanisation in the northern parts of the Nordic area, combined with the modernisation of the industry' (BIN 2017,45). On the other hand, this has resulted in an increased demand for high-skill labour and an increase of job creation in the real estate, professional, scientific, and technical sector (1,984 jobs), followed by public administration, defence, and social security since 2008 (BIN 2017, 45). Analysis has also shown that the population in the Barents Region is generally ageing and becoming more urbanised, which has important economic and social consequences (AMAP 2017, 6). There is indeed important variation in unemployment and education rate between urban and non-urban areas (BEAC 2015, 15).

The level of prosperity also largely differs between the Russian and Nordic parts of the Barents Region. With the dismantling of the USSR, unemployment has increased, and a wider gap between rich and poor has appeared in Russia and its Barents Regions (Bahry, 2002; Duhaime and Caron 2008; Rosstat in Didyk 2012, 148). The poverty rate is reported to be between 20% and 25%, affecting mostly single parent families and large families, and individuals with little education and those living in rural areas (Bahry, 2002; Duhaime and Caron 2008). In all of the regions of the Russian part of the Barents, the threshold value of the decile ratio, which indicates income inequality, was exceeded to varying degrees between 2003 and 2008 (Didyk 2012, 148). The changing role of the Russian state with the withdrawal of the government as producer and organiser 'have redrawn class boundaries, undermined traditional job guarantees and eroded the old social safety net' (Bahry, 2002; Duhaime and Caron 2008). Since the economic transition, the Russian Barents is also facing demographic challenges caused by a massive migration from the Arctic regions to the southern regions (Eikeland & Riabova 2002). Life expectancy is also critically low. According to 2008 census, 'in none of the Barents Russian regions did life expectancy reach the "threshold" level used as a criterion for estimations of sustainability' and 'none of the BEAR regions reached the national average level of life expectancy' (Didyk 2012, 146). In fact, while all Russian regions of the Barents have improved their HDI during the period from 2003 to 2008, the index values are still below the national average and in Murmansk Oblast and Karelia the values are below the level which is considered the lower level for developed countries (0.8) (Didyk 2012, 145–146). As indicated by Duhaime and Caron (2008, 20), 'the socioeconomic conditions of the "New Russia", built on the Soviet ruins, thus reflect a greatly weakened social situation and a redistribution system which cannot even be qualified as a rudimentary welfare state'.

By contrast, the Nordic countries have achieved some of the most favourable social conditions for human development in the Arctic Region (Duhaime and Caron 2008, 18). Poverty rates in Fenno-Scandinavia are lower than elsewhere, life expectancy and education level are the highest within the circumpolar Arctic, and infant mortality is the lowest (Glomsrød and Aslaksen 2009, 17–18). As a result, all three countries are ranked within the top 25 countries in the world HDI ranking (UN 2017). This is partly due to the social democratic tradition and the political redistribution of resources through social policies. All indicators show that redistribution policies have had beneficial effects on living conditions in the Arctic regions of the Fenno-Scandinavian countries (Glomsrød and Aslaksen 2009, 17–18). However, all regions of the Barents area are experiencing an ageing of its population, where the proportions of adults and elderly increase, while the proportions of children and adolescents decrease. This process results in a rise in the median age of the population and an increase in the dependence ratio which 'indicates growing pressures on social security and public health systems' (BIN 2017, 11). This situation can lead to serious problems in the economic and social sphere and could affect economic security in this part of the region as well. Furthermore, there are also income disparities that have been observed between the Sámi and other inhabitants. While the Sami people are certainly less impoverished than indigenous peoples in the Russian parts, it has been observed that the income for the area where Sami live in Norway is considerably lower than the average total household income for other northern areas (Glomsrød and Aslaksen 2009, 116–117). Additionally, it is also reported that Sámi living in Finland 'earn less than other inhabitants of the country and are under-represented on the job market and over-represented among the unemployed' (Duhaime and Caron 2008, 19). Thus, economic security varies between countries, east from west, and between peoples.

Beyond those differences, it can be noted that one of the common challenges faced by the Barents population is the development of sustainable and environmentally responsible economic activities (BEAC 2015, 16). Whereas extractive industries play a major role in the economy of the region, the pristine environment and its renewable resources are major sources of livelihoods for the Barents population in all parts of the region. Both the tourism industry, which is one of the most important sources of income in the Barents Region, and local and indigenous communities livelihoods depend on the sustainable management of the Barents environment. However, an increased demand for non-renewable resources in nature based activities has exacerbated existing conflicts between extractive and renewable industries, more particularly increasing tension on traditional hunting, fishing, and herding by indigenous peoples (BIN 2017, 8). Finally, the Barents economy is influenced by global development, whether through migration, trade, or geopolitical events (AMAP 2017). Although it is not possible to analyse or predict how this development may influence the region in the future, those are elements that also impact the state of economic security of the Barents population.

Considering these contexts and issues, the following section assesses different aspects of economic (in)security that challenge the contemporary wellbeing of the Arctic population, focusing especially on the Barents Region.

3 Assessment

Economic security encapsulates complex and interlinked challenges, which highlight economic development as a means for achieving a larger sense of security, sustainability and wellbeing for peoples. This section focuses on some of the main economic security challenges faced by the population in the Barents Region. These diverse challenges include an assured access to employment, education, demography changes, transportation systems, sustainable development and the impact of mining industries on subsistence economies as well as the external effects of geopolitical and economic events on the region.

3.1 Unemployment

As already mentioned above, economic indicators reveal differences across countries in the Barents Region. One of the relevant indicators is unemployment, which is also one of the most commonly used tools to measure economic security. One of the major effects of unemployment is indeed its potential to drive individuals into poverty.

Unemployment rates vary significantly within the Barents Region, generally with Troms (Norway) having the lowest and Kainuu (Finland) having the highest ratio. In 2014, Troms had an unemployment rate of 1.6%, while the ratio was 16.9% in Kainuu. The unemployment rate in other regions were as follows: 1.9% in Nordland (Norway), 2.4% in Finnmark (Norway), 7% in Västerbotten (Sweden), 7.7% in Norrbotten (Sweden), 9.7% in Lapland (Finland), 10.2% in Northern Ostrobothnia (Finland), 6.7% in Murmansk (Russia), 8.1% in Karelia (Russia), 6.2% in Arkhangelsk (Russia), 6.1% in Komi Republic (Russia), and 6.3% in the Nenets Autonomous Okrug (BEAC 2015, 15). Generally, the unemployment rate in the Norwegian part of the Barents Region was lower than the national average, while in the other three Barents countries, the unemployment rate in the Barents Region was higher than in other parts of the countries. As the above figures show, the unemployment rate in different parts of the Barents Region vary from approximately 1.6% to 16.9%, but there is no great discrepancy in the region overall.

As already mentioned, it has also been observed that the biggest employment losses which occurred in the Nordic part of the Barents Region concerns mining, quarrying, and manufacturing, followed by agriculture, forestry, and fishing (BIN 2017, 38). This trend reflects a decline in the employment share of middle-skilled and middle-waged occupations partly caused by growing automation in production, job outsourcing, and price competition from emerging countries, all which affect manufacturing jobs (BIN 2017, 38). Furthermore, there is also a high unemployment rate among young people

(BEAC 2014, 2). As a consequence, and in order to increase economic security, an up to date assessment of the labour market situation and needs of the Barents population is necessary. Further research is also needed on how to support more favourable employment development in the overall region and increase new employment opportunities.

3.2 Demographic changes

The Barents Region faces significant demographic challenges due to its increasingly ageing population, and the trend of younger populations leaving the Barents Region to work in larger cities in the south of the respective countries (BEAC 2015, 5). Europe in general is currently the oldest continent with the highest old age dependency ratio, and according to projections it will remain so until 2060 (EU Commission 2014, 23). The implications of these demographic changes are even more severe for the Barents Region. The falling population trend is caused by negative net migration, and declining birth rate. Young and highly educated people are migrating mostly from peripheral towns to urban centres in the south. In the past decades the region has lost thousands of working-age citizens due to a lack of job opportunities, especially in the Russian regions and in the Kainuu region in Finland (BEAC 2015, 12).

Since 1990, the population has been declining in the Barents Region, especially in the Russian part. The greatest population decline is observed in Murmansk – in 2014, statistics showed a 35.29% decline compared to 1990 (Patchworkbarents in BEAC 2015, 13). In Arkhangelsk Oblast, the population between 1990 and 2015 changed from 1,575,502 to 1,183,323. The Komi Republic and the Republic of Karelia also experienced significant population decline within the same period. The population in the Komi Republic changed from 1,248,891 to 864,424, in the Murmansk Oblast from 1,191,468 to 766,300, whereas in Karelia from 791,719 to 632,500 (Patchworkbarents in BEAC 2015, 13). In other parts of the Barents Region, the demography changes have not been as harsh. Within 1990 and 2015, the population of the region has changed as shown in Table 2.2.1 (indicated in descending order in terms of population).

Although in some areas (for instance in Northern Ostrobothnia in Finland, Västerbotten in Sweden, as well as throughout the Norwegian Barents) statistics showed an increasing population between 1990 and 2014 (Patchworkbarents in BEAC 2015, 13), the total population in the Barents Region within that period still showed a decline of at least 20.93% (BEAC 2015, 13). The general decline in population is largely due to the declining population in the Russian areas of the Barents Region. Furthermore, even if the Nordic part of the Barents have experienced a population growth, the declining population in the age group 0–19 and considerable increase in population for the age group 65+ may have long term implications for labour and education (BIN 2017).

Table 2.2.1 Demographic trends in the Barents Region

	1990	*2015*
Northern Ostrobothnia	346,301	406,966
Norrbotten	263,735	249,777
Västerbotten	251,968	262,593
Nordland	239,532	241,682
Lapland	200,674	180,848
Troms	146,594	163,453
Kainuu	96,957	78,388
Finnmark	74,148	75,605

Source: data from Patchworkbarents (BEAC 2015, 13)

The implications of these demographic changes are also important for transfer systems and the ratio of economic dependency. For instance, the Nordic parts of the Barents Region are experiencing an ageing of its population, where the proportions of adults and elderly increase, while the proportions of the youth decrease. Population in the age class 0–19 in the Nordic parts of the Barents decreased by 5.9 % while for the national average as a whole it grew by 1.9 % in the 2006–2015 period (BIN 2017, 18). Statistics also shows that in 2015 half of this area's population was older than 41.8 years, while the other half was younger. However, with a rise in the dependency ratio, there is a growing pressure on social security and public health systems in the region. In this context, important political and development measures should be taken, especially concerning the role of the elderly population as participants in the Barents economy (BIN 2017, 21).

3.3 Education needs and resources

A lack of access to quality higher education can also manifest in economic poverty and unemployment, resulting in insufficient economic resources. Conversely, prior studies indicate that highly educated peoples generally have better health and have higher employment rates and higher relative earnings (OECD, 2016 in BIN 2017, 36). Tertiary education also contributes to foster innovation, increases economic activities and growth, and contributes to the wellbeing of the population. In the Nordic countries, a growth in tertiary education has been observed but the percentage of the population aged 20–59 who have successfully completed tertiary studies still lag behind the national average (BIN 2017). This statistic indicates that policy actions could be envisaged to improve tertiary education in the Nordic countries.

In addition, it is also necessary to ensure proper primary and secondary education in the Barents Region. However, ensuring basic and adequate education is challenging for communities inhabiting remote areas. The most significant challenges are the consolidation and closing of small, remote schools, providing sufficiently comprehensive education opportunities, as well as recruiting and retaining teachers in remote communities (Larsen and Fondahl 2015, 351). Shrinking populations in remote parts of the Barents also constitute a challenge to maintaining access to education services (BIN 2017, 37). Another important factor includes language instruction, especially with regard to the efforts to retain traditional languages (i.e. languages other than the dominant language in a given society, such as for example Sámi languages) and to ensure that students acquire the national languages of commerce and higher education. In addition, finding new ways to use and transmit indigenous knowledge, the gender pay gap, as well as the continuing underperformance of indigenous students across most parts of the North are all relevant social issues that influence individuals' access to economic opportunities (Larsen and Fondahl 2015, 351).

3.4 Transportation and infrastructure

Interaction between citizens, businesses, and other civil society actors is an important factor for broader economic growth and economic security. In the Barents Region, this interaction is intrinsically dependant on adequate transportation and infrastructure systems. Moreover due to the fact that population density in the Barents Region is low, and the main markets for raw materials and refined products are located outside the region, transport and infrastructure have significant importance in the Barents. The significance of transportation is also reflected in other aspects, such as its impact on food costs, business development, and cultural relationships between peoples and communities.

Specifically, large quantities of time-sensitive seafood are delivered to both western and eastern Europe, but transport is also important for other sectors of the economy. Raw materials and manufactured products from the mining, metal, and forest industries also need to be transported to Europe for further processing. Asia and America are also important markets for raw materials (BEAC Transport Area 2013, 37). Some raw materials and manufactured products are also traded within the Barents Region. For instance, ore from Russia and Sweden are transported for processing in Finland, and seafood from Northern Norway is delivered across the region. However, the largest intra-Barents flows are transit flows: for instance, half of the farmed salmon produced in Northern Norway passes through Northern Sweden and/or Northern Finland on its way to the market. For example, farmed salmon from Northern Norway is transported by road to Helsinki, where it is made ready for further transport by plane from Helsinki to Asia for distribution. Also, iron ore from Northern Sweden transits Northern Norway on its way to the biggest markets. Oil and gas extracted in the Russian Federation is transported along the coast of Norway, but mainly through international waters. Furthermore, there are a number of traffic hubs in the region where re-loading between different types of traffic can take place. In larger places there are strategic cargo nodes, which are intermodal, whereas in other places there are railway stockyards or port facilities. (BEAC Transport Area 2013, 37–38).

The importance of transportation in the region is therefore evident. The existence of a strong transportation infrastructure is not only crucial for the rapidly expanding industries and for the development of remote areas, but also contributes to economic stability and mobility of labour. However, interaction between citizens, businesses, and other civil society actors in the Barents Region is hampered by inadequate transportation and infrastructure systems (BEAC 2015, 16). While there is relative good transportation in a North–South direction, the lack of transport infrastructure East–West significantly hinders economic development across the borders of the Barents Region. In order to ensure smooth transportation and border crossings, well-functioning cross-border cooperation in the Barents Region is consequently required. More specifically developing East–West transport infrastructure is a key to business cooperation in the Barents Region and 'further extension of the traffic infrastructure eastwards appears to be vital' (BIN 2017, 93).

3.5 Economic and sustainable development

The economic security and wellbeing of the population depend in part on the living environment and natural resources of the Barents Region. Regional development needs to happen in concordance with preserving the environment, mitigating and adapting to climate change, and fostering good and healthy living conditions for the people. The attractiveness of the region in enticing people to live and work in the Barents Region, as well as for the establishment of businesses, is relevant to subsequent employment opportunities, education, health, culture, the surroundings, nature, and recreational activities (BEAC 2014, 4).

As a consequence of the growing global need for minerals and other natural resources, extractive industries are continuously expanding. As the Barents has already proven to be rich in natural resources, extractive industrial activities have become important contributors to the national economies of states. The development of national legislation and mineral strategies reflects the interest of states in the expanding potential of mining industries. For instance, the vast majority of the iron ore produced in the EU comes from Sweden, and the Swedish government has expressed its wish to strengthen its position as the leading mining nation in the EU (Swedish Ministry of Enterprise, Energy and Communications 2013). Furthermore, Finland is a global pioneer in the responsible use of minerals, and the country's legislation is amongst the most attractive for mining investments worldwide. The Norwegian government defined its mineral industry as a focal area, stating that it wants the country to be attractive for mining activities (Norwegian Ministry of Trade and Industry 2013). The Russian Federation is also exceptionally rich in natural resources, and subsoil resources significantly contribute to the country's economy (Pettersson et al. 2015, 247).

As already mentioned, extractive industries play an important role in the economy of the region. However, the mining industry and other extractive development projects may have an adverse impact on the region's nature and environment, which may undermine environmental sustainability. While the reasonable and economic acquisition and use of mineral resources is gaining more and more significance (Dubiński 2013, 2), the impact of extractive industries in the Barents nevertheless constitutes a challenge for the Barents environment and other economic activities. This is the case with tourism, a main driver of economic growth in the Nordic countries and an integral part of the local economy in the Barents Region (AMAP 2017, 12). The construction of mining sites may damage the natural environment near the mine, which can be problematic as pristine nature is one of the most attractive factors for the tourism industry in the Barents Region. Ultimately, one of the main challenges faced by the region is therefore to ensure an adequate balance between competing interests and actors to ensure that they contribute to the economic, social, and environmental sustainable development of the entire population and communities living in the region.

Another challenge faced by the Barents population relates to the revenue generated by extractive industries and their impact on local population. Economic benefits from the activities of extractives industries are not always used for the benefit and interest of the local population. Despite the significant revenue generated by extractives industries in the Russian region, the share of the population living in this part of the region with income below subsistence level is critically high. In 2008, only in the Nenets Autonomous Okrug (AO) did the poverty level indicator 'not exceed the "threshold" level when estimated in terms of the criteria for sustainable development'. In all other regions, the poverty level was considered critically high and incompatible with sustainable development (Didyk 2012, 147). Furthermore, although the poverty level of the Nenets AO is not considered critical, the level of income inequality is the highest in Russia. This is explained by the influence of extractive industries and simultaneous presence of low-income population living in this region (Didyk 2012, 148). This situation clearly demonstrates the importance of redistribution policies to alleviate income disparities, across the Barents, which should evidently been promoted in Russia.

Finally, another challenge concerns the impact of economic development on the subsistence economies of indigenous peoples. The Barents Region is home to several indigenous peoples, which also indicates different economic livelihoods. The livelihoods of indigenous peoples typically relate to land, water, and other natural resources. In the Barents Region, indigenous peoples have developed highly specialised livelihood strategies and occupations that include, among others, reindeer herding, hunting, fishing, trapping, shifting cultivation or gathering food and forest products, and, in some cases, handicrafts. They are therefore dependent on the rights to natural resources and the management of natural resources for their subsistence. Because of indigenous peoples' dependence on natural resources, their interest in preserving these resources in the long term is significant (Skogvang 2013). These livelihoods are not only a source of food and income for indigenous peoples, but are also a part of their heritage and culture.

Problematically, indigenous peoples's traditional livelihoods also compete with other societal interests (Allard 2006, 15), more particularly the need to increase economic development via the development of resource extractive projects. However, the extractive industries in the Barents Region are posing a real threat to the livelihoods of the indigenous peoples, such as the Sámi and the Nenets (Anaya 2011, para. 55; Tauli-Corpuz 2016, para. 86). In many cases, development projects represent a basis for conflicts between industries and the affected indigenous peoples. For instance, an important Sámi protest took place in Jokkmokk (Sweden) in 2013, when test mining permission was granted to a British mining company on reindeer grazing lands (Koivurova et al. 2015, 28). In addition, extractive industries also threaten the traditional way of life of indigenous peoples in the Russian Federation, such as the nomadic lifestyle (including hunting, fishing, and reindeer herding) of the Nenets people (Garipov 2014, 74–75; Anaya 2010).

Behind the impact of extractive industries, arctic indigenous peoples, such as the Sámi and Nenets, face other challenges due to environmental problems, for instance pollution and climate change, which seriously affect their rights to their traditional territories (Heinämäki 2004, 231–233). Ultimately, in order to ensure sustainable development for all in the Barents Region, policy actions should recognise and implement the rights of indigenous peoples in the region in a better fashion. Specifically, measures to include them in the decision-making processes concerning the development of their traditional land and territories should be promoted in order to ensure that their voice and interests are represented in the governance of their territories (See also chapter 2.8).

3.6 External influences of geopolitics and financial events

Economic security is also influenced by economic and geopolitical events. External economic and foreign policy instability can be responsible for unstable market and economic relations, which can affect the economic security of peoples at the local level. The 2008 global economic crisis has certainly influenced the Barents market. For instance, the impact on mineral and ore prices negatively influenced northern counties in both Sweden and Finland (BIN 2017, 91). Similarly, the Russian economy has been deeply affected by the fall of the global oil prices, leading to significant revenue shortfalls and recession in the country. As a result of the recession, Russia is reversing substantial achievement in poverty reduction. In 2015, the poverty rate in the country increased from 11.2 to 13.4%, as the poor population increased by 3.1 million to a total of 19.2 million (World Bank Group 2016, 8). In 2016, a further increase in poverty levels was projected, due to a continuing increase in unemployment and government's difficulty in enforcing poverty-reduction measures (World Bank Group 2016, 8). Thus, the Russian economy faces important challenges that are both induced by internal and external phenomena such the global financial crisis and lower oil prices. Although it is not defined how this trend has distinctively affected the Barents Region, there is no doubt that these developments have also negatively impacted its population.

In relation to the impact of external events on economic security, the consequences of the Ukrainian crisis on the Russian and Barents economy is another case in point. Following Russia's annexation of Crimea, the European Union responded with a series of sanctions, including asset freezes on key allies of President Putin, an arms embargo, restrictions on access to capital markets, and several other targeted measures. Furthermore, these measures were not limited to, or principally focused on the occupied territory, but were explicitly adopted as diplomatic tools in response to Russian actions. These sanctions had consequences for the Barents Region as well.

As a consequence of EU sanctions, Lapland, Northern Ostrobothnia, and Kainuu experienced a lower demand for paper and shrinking exports to Russia as well as reduced tourism inflows (BIN 2017, 8). Within the Barents Region, Finland and Norway have been disproportionately affected by Russian counter-sanctions on agricultural, livestock, and fishery products in comparison to other European countries (Larrabee et al. 2017, 27). Dairy farms were the most affected by counter-sanctions, and some were even forced to close down (Nilsen 2016). In addition, even though tourism in the Russian Barents had already been declining, the sector witnessed a 3% drop by May 2014, reinforced by the political crisis (Nilsen 2014). As part of the sanctions, the defence industry struggled to replace Western arms imports, and energy companies had to deal with restrictions on energy cooperation with Russia. Furthermore, stability in business and banking sectors had to be managed due to financial restrictions (Russel 2016, 7–9). Finally, the 2014 crisis and subsequent sanctions have also had an 'impact on EU-Russia cooperation in many sectors and have blocked several of the previously available multilateral financing sources to foster the Barents cooperation' (BEAC 2015, 7).

In light of this, one important recommendation therefore lies in the need to enhance and strengthen economic cooperation across borders. Since administrative and political cooperation between members at the regional level of the Barents has continued and become stronger after the annexation of Crimea (Koivurova 2016), it is a primary concern to reinforce economic bonds in the region, a task that can be channelled via existing institutions. Enhancing the level of funding both for large scale and small-scale cross-border activities, and businesses, could also contribute to such a development, especially as 'the majority of Barents Regions consider the current level of financing insufficient for their joint activities' (BEAC 2015, 7). Among other actions, addressing structural constraints to investment in Russia would also help enhance long-term growth prospects (World Bank Group 2016, 8) and perhaps alleviate economic insecurity for the population. Thus, improving economic security requires a bundle of measures to be taken at the regional, national, and international levels.

4 Conclusion

Projections show that the world economy will grow at around 3% per year over the next fifty years (although according to recent forecasts, this number will go down) but the OECD projects that the Barents countries will see economic growth rates below the world average. Nonetheless, all four countries will roughly double the size of their economies by 2050, with the highest real GDP growth in Sweden (110%) and the lowest in Russia (93%) (AMAP 2017, 11). However, the Barents Region is currently facing various challenges in terms of their economies. This includes demographic changes and outmigration from the Barents Region, higher unemployment, lack of access to quality higher education in remote communities, hindered transportation possibilities, income inequality, conflicts between indigenous and local communities vis-à-vis expanding mining and energy industries, as well as the collateral effect of the external events such as the financial crisis of the Ukrainian conflict.

There is certainly not an all-encompassing solution to these issues. Nonetheless, local action plans accompanied by a strong regional co-operation are required from nation states in order to improve the economic security of the population. Attractive living environments should also be promoted in the Barents Region in order to get people to move in, stay, or return to the region (BEAC 2014, 2). More precisely, there is an increasing need to create new employment possibilities, since unemployment, especially among young people, is quite high in some parts of the region. There is also a need to redefine the role of the elderly as active participants in the Barents economy (BIN 2017, 21). While access to better primary and secondary education must be improved, it is also important to support tertiary education in order to stimulate innovation and economic development in the region. Furthermore, transportation routes and infrastructure need to be developed throughout the region. Finally, innovation and industrial development must be enhanced, especially in the field of renewable nature-based industries in order to support sustainable economic development in the region. However, the benefits of development should be distributed to the population in a manner that benefits their interests at the local level and the protection of indigenous peoples' rights needs also to be guaranteed – a challenging task that will necessarily require policy and legislative reforms. Lastly, the region would benefit from stronger cooperation and further funding to support cross borders activities.

Note

1 Employed people are those aged 15 or over who report that they have worked in gainful employment for at least one hour in the previous week or who had a job but were absent from work during the reference week. The working age population refers to people aged 15 to 64.

References

Allard, C. 2006. *Two Sides of the Coin: Rights and Duties – The Interface Between Environmental Law and Saami Law Based on a Comparison with Aoteoaroa/New Zealand and Canada*. Sweden: Luleå.

AHDR 2004. *Arctic Human Development Report*. Akureyri, Iceland: Stefansson Arctic Institute.

AMAP (Arctic Monitoring and Assessment Programme). 2017. *Barents Area Overview Report*. Oslo, Norway: Adaptation Actions for a Changing Arctic (AACA) – Barents Area Overview report. www.amap.no/documents/doc/Adaptation-Actions-for-a-Changing-Arctic-AACA-Barents-Area-Overview-report/1529.

Anaya, J. 2010. *Report of the Special Rapporteur on the Situation of Human Rights and Fundamental Freedoms of Indigenous People – Situation of Indigenous Peoples in the Russian Federation*. A /HRC/15/37/Add.5. Human Rights Council. http://undocs.org/A/HRC/15/37/Add.5.

Anaya, J. 2011. *Report on Extractive Industries Operating within or near Indigenous Territories*. A/HRC/18/35. Human Rights Council. http://www.ohchr.org/Documents/Issues/IPeoples/SR/A-HRC-18-35_en.pdf.

BEAC Transport Area. 2013. *Joint Barents Transport Plan – Proposals for Development of Transport Corridors for Further Studies*. www.barentsinfo.fi/beac/docs/Joint_Barents_Transport_Plan_2013.pdf.

BEAC. 2014. *The Barents Programme 2014–2018*. www.barentsinfo.fi/beac/docs/Barents_Programme_2014_2018_adopted_2_June_2013.pdf.

BEAC. 2015. *Financing of Barents Cooperation – Report of the BEAC Ad Hoc Working Group on Financial Mechanism Study*. Ministry for Foreign Affairs of Finland. www.barentsinfo.fi/beac/docs/LOW_UM_Barents_eJulkaisu_A5.pdf.

Bahry, D. 2002. Ethnicity and Equality in Post-communist Economic Transition: Evidence from Russia's Republics. *Europe-Asia Studies* 54(5): 673–699.

Business Index North (BIN), Bullvåg, E.*et al.*2017. *A Periodic Report with Insight to Business Activity and Opportunities in the Arctic*. http://www.kpb.no/sites/k/kpb.no/files/businessindexnorth2017web.pdf

Dubiński, J. 2013. Sustainable Development of Mining Mineral Resources. *Journal of Sustainable Mining* 12(1): 1–6.

Duhaime, G. and Caron, A. 2009. Economic and Social Conditions of Arctic Regions 2008. In *The Economy of the North*, edited by S. Glomsrød and I. Aslaksen. Oslo: Statistics Norway.

Didyk, V. 2012. Sustainable Development and Local Self-government in the Russian Part of the Barents Region. In *Politics of Development in the Barents Region*, edited by M. Tennberg. Rovaniemi: LUP.

Eikeland, S. and Riabova, L. 2002. Transition in a Cold Climate: Management Regimes and Rural Marginalisation in Northwest Russia. *Sociologia Ruralis* 42(3): 255–266.

EU Commission. 2014. *The 2015 Ageing Report – Underlying Assumptions and Projection Methodologies*. European Economy 8|2014. European Economy Series. http://ec.europa.eu/economy_finance/publications/european_economy/2014/pdf/ee8_en.pdf.

Garipov, R. 2014. Extractive Industries and Indigenous Minority Peoples' Rights in Russia. *Nordic Environmental Law Journal* 1: 67–75.

Gentilini, U. and Omamo, W. 2009. Unveiling Social Safety Nets. *Occasional Paper* No 20. World Food Programme. https://www.wfp.org/sites/default/files/OP20%20-%20Unveiling%20Social%20Safety%20Nets%20-%20English.pdf.

Glomsrød, S. and Aslaksen, L. 2009. *The Economy of the North 2008*. Oslo: Statistics Norway.

Heinämäki, L. 2004. Environmental Rights Protecting the Way of Life of Arctic Indigenous Peoples: ILO Convention No. 169 and the UN Draft Declaration on Indigenous Peoples. In *Arctic Governance*, edited by T. Koivurova, T. Joona and R. Shrono. Rovaniemi: Juridica Lapponica 29.

ILO. 2004. *ILO Socio-Economic Security Programme*. www.ilo.org/public/english/protection/ses/download/docs/definition.pdf.

Koivurova, T. 2016. Analysis: The Arctic Conflict – Truth, Fantasy or a Little Bit of Both? *High North News*, November 18. www.highnorthnews.com/analysis-the-arctic-conflict-truth-fantasy-or-a-little-bit-of-both/.

Koivurova, T, Masloboev, V., Hossain, K., Nygaard, V., Petrétei, A. and Vinogradova, S. 2015. Legal Protection of Sami Traditional Livelihoods from the Adverse Impacts of Mining: A Comparison of the Level of Protection Enjoyed by Sami in Their Four Home States. *Arctic Review on Law and Politics* 6(1): 11–51.

Larrabee, F.S., Pezard, S., Radin, A., Chandler, N., Crane, K. and Szayna, T. 2017. Russia and the West after the Ukrainian Crisis: European Vulnerabilities to Russian Pressures. RAND Corporation. www.rand.org/content/dam/rand/pubs/research_reports/RR1300/RR1305/RAND_RR1305.pdf.

Larsen, J. and Fondahl, G. 2015. *Arctic Human Development Report*. Copenhagen: Nordic Council of Ministers.

Nilsen, T. 2014. Ukraine Crisis Hurts Russian Tourism. *Barents Observer*, May 8. http://barentsobserver.com/en/business/2014/05/ukraine-crisis-hurts-russian-tourism-08-05.

Nilsen, T. 2016. EU Has No Plans to Lift Russia Sanctions. *The Independent Barents Observer*, November 16. https://thebarentsobserver.com/en/life-and-public/2016/11/eu-has-no-plans-lift-russia-sanctions.

Norwegian Ministry of Trade and Industry. 2013. Strategy for the Mineral Industry. www.regjeringen.no/contentassets/3fe548d142cd496ebb7230a54e71ae1a/strategyforthemineralindustry_2013.pdf.

Pettersson, M., Oksanen, A., Mingaleva, T., Petrov, V. and Masloboev, V. 2015. License to Mine: A Comparison of the Scope of the Environmental Assessment in Sweden, Finland and Russia. *Natural Resources* 6: 237–255.

Russel, M. 2016. Sanctions over Ukraine: Impact on Russia. PE 579.084. European Parliamentary Research Service. European Parliament. www.europarl.europa.eu/EPRS/EPRS-Briefing-579084-Sanctions-over-Ukraine-impact-Russia-FINAL.pdf.

Sen, A. 1983. Development: Which Way Now? *The Economic Journal* 93(372): 745–762.

Skogvang, S. F. 2013. Legal Questions Regarding Mineral Exploration and Exploitation in Indigenous Areas. *Michigan State University College of Law Journal of International Law* 22(1): 321–345.

Swedish Ministry of Enterprise, Energy and Communications. 2013. *Sweden's Minerals Strategy – For Sustainable Use of Sweden's Mineral Resources That Creates Growth throughout the Country*. http://www.government.se/49b757/contentassets/78bb6c6324bf43158d7c153ebf2a4611/swedens-minerals-strategy.-for-sustainable-use-of-swedens-mineral-resources-that-creates-growth-throughout-the-country-complete-version.

Tauli-Corpuz, V. 2016. *Report of the Special Rapporteur on the Rights of Indigenous Peoples on the Human Rights Situation of the Sami People in the Sápmi Region of Norway, Sweden and Finland*. A /HRC/ 3 3 / 4 2 /Add. 3. United Nations General

Assembly. http://unsr.vtaulicorpuz.org/site/images/docs/country/2016-sapmi-a-hrc-33-42-add-3-en.pdf.
United Nations. 2017. *Human Development Index*. http://hdr.undp.org/en/content/human-development-index-hdi
UNDP. 1994. *Human Development Report*. New York: Oxford University Press.
UNDP. 2016. *Human Development Report 2016 – Human Development for Everyone*. www.hdr.undp.org/sites/default/files/2016_human_development_report.pdf.
World Bank Group. 2016. *Russia Economic Report No. 35 I April 2016 The Long Journey to Recovery*. http://pubdocs.worldbank.org/en/481881460390188506/rer35-ENG.pdf.

2.3 Health security in the Barents Region

Susanna Pääkkölä and Dorothée Cambou

Health security constitutes a fundamental component of human security. The main goal of the present chapter is to highlight some key challenges to health and wellbeing in the Barents Region.

1 Definition

One of the most significant human achievements of the 20th century was progress in health and wellbeing (Commission on Human Security 2003). 'About a billion people today have average life expectancies of nearly 80 years, twice the average of a century before'. Yet, good health remains inequitably distributed (Commission on Human Security 2003, 95) and health security has acquired a renewed salience in the human security discourse (Elbe 2010, 101).

According to the Commission on Human Security, 'good health is both essential and instrumental to achieving human security' (2003, 96) because the very purpose of security is protecting human lives (Elbe 2010, 101). In this regard, health security is linked to unnecessary deaths that can be prevented by better health behaviour or by reaching people trapped in poverty or conflict (Commission on Human Security 2003, 95). The Commission on Human Security identifies global infectious diseases, poverty-related threats, and violence and crisis as three major health challenges. However, health is not only defined the absence of disease or death. According to the Commission on Human Security, health also includes 'a state of complete physical, mental and social well-being'. It integrates mental health and physical heath as well as subjective psychosocial wellbeing and confidence about the future, which results in a more holistic approach to disease prevention and health promotion (Commission on Human Security 2003, 95).

Therefore, assessing health security goes beyond measuring morbidity and mortality. It includes an assessment of how people perceive their own health and the trajectory of their lives. The achievement of health security also takes into account interrelated issues including pre-existing health, work, economic and family situations. For instance, it takes into consideration that higher living standards are associated with decreased risks of disease, illness and injury, better immune functioning, speedier recovery and increased longevity. It also considers factors that increase life expectancy such as good nutrition, physical exercise and safe sex. Finally, it also includes psychological wellness and the capacity for 'people to exercise choice, pursue social opportunities and plan for their future' (Commission on Human Security 2003, 96).

At the international level, there are different indicators to evaluate health security. Under the United Nations Indicators of Sustainable Development (United Nations 2007, 11), which serves as reference for countries to develop or revise national indicators of sustainable development, health security is determined through four core indicators:

1 Under-five mortality rate (life expectancy at birth and healthy life expectancy at birth),
2 Health care delivery (percentage of population with access to primary health care facilities, contraceptive prevalence rate and immunisation against infectious childhood diseases),
3 Nutritional status (nutritional status of children) and
4 Health status and risks (morbidity of major diseases such as HIV/AIDS, malaria, tuberculosis and prevalence of tobacco use).

While these indicators are helpful in assessing a quantitative perspective of health security, they are limited in their ability to measure what people think and feel about their lives. The individual wellbeing of peoples is often expressed in relative terms, and is also connected to personal judgements of life satisfaction and feelings ranging from depression to joy. Access to healthcare and services to promote wellbeing are also nuanced by specific individual needs associated with gender, sexual orientation, age, culture and disability. Furthermore, while health insecurity concerns us all, certain vulnerable groups of people are disproportionally affected. In its report on human security, UNDP highlights that 'in both developing and industrial countries, the threats to health security are usually greater for the poorest, people in the rural areas and particularly children' (UNDP 1994, 28). In this regard, health insecurities are relative and dependent on the infrastructure and demographics of the region, and therefore must be contextualised. Finally, in the age of global migration, health concerns are no longer isolated national issues, but require regional and cross-border considerations. This situation also calls for an assessment of the specific needs of minorities and indigenous populations, and communities beyond national majorities.

2 Contextualisation

In the Barents, challenges to healthcare differ substantially in each respective country.

> In both Finland, Norway, Sweden and the Russian part of the region, the organisation of the health service is affected by the fact that this is a very

> sparsely populated area with long distances, for which reason the health service provided differs significantly from the health service provided in the more densely populated areas of the countries.
>
> (JWGHS Statistic Working Group 1998)

In addition, population demographics and vital statistics between the three Nordic countries and Russia vary substantially. Whereas the population in the Finnish, Norwegian and Swedish parts of the Barents Region share close similarities, the population in the Russian part differs significantly (JWGHS Statistic Working Group 1998).

In particular, mortality rates vary significantly in the Barents Region. Fennoscandia has the lowest mortality rate in the world, whereas in the Russian Barents Regions the rates are relatively high. In 2014, mortality rates varied from around 8% (total deaths) in the North Ostrobothnia/Troms regions to roughly 14% in Karelia and Arkhangelsk (Emelyanova and Rautio 2016, 7). In addition, significant cross-regional differences in mortality are shown in the differences in life expectancy at birth for males/females. First, life expectancy for the men living in regions belonging to the Nordic countries is higher than for those living in Russia. In 1998, it was noted that men from the Nordic countries could expect to live 12 to 15 years longer than their Russian counterparts, whereas the women in the Nordic countries could expect to live around 10 years longer than their Russian counterparts (JWGHS Statistic Working Group 1998). In addition, 'in the Norwegian, Swedish and Finnish regions, women live significantly longer than men' (BEAC 1998). This data has not seen significant changes in the past two decades. In 2014, it was noted that life expectancy was 63 for women and 76 for men in Karelia, compared to 80 for men and 83 for women in Vesterbotten (Emelyanova and Rautio 2016, 7). Finally, the respective regional life expectancies of both men and women in the Barents Regions (the northern areas of each country) are slightly lower than the respective national averages in Norway, Sweden and Finland.

Beyond mortality rates, diseases patterns and causes of death vary across the region. Twenty years ago,

> prostate cancer and cancer of the bladder were most widespread in the Norwegian and Finnish regions, whereas cancer of the stomach was most widespread in the Russian regions. The same applies to cancer of the respiratory organs, which was also very high for the Finnish region. Cancer of the mouth and esophagus was most prevalent in the Murmansk and Archangelsk regions.
>
> (JWGHS Statistic Working Group 1998)

In 2016, reported causes of death in Barents Region have changed. Currently, the most common are cardiovascular diseases (Emelyanova and Rautio 2016, 7). Russia leads with the highest number of deaths, resulting from general heart diseases but also from malignant neoplasms. In comparison, the situation is different in the Troms and Finnmark regions, where respiratory diseases as well as accidents and violent deaths are the main causes of mortality (Emelyanova and Rautio 2016, 8). But, these are quite rare in Sweden, where the highest cause of disease-related death is provoked by cancer, especially lung cancer for men and breast cancer for woman. A similar situation exists in Finland, where alcohol poisoning is also a major factor of disease-related deaths (Emelyanova and Rautio 2016, 8).

In the Barents, indigenous peoples, Sami, Nenets and Vepsians are also the most vulnerable peoples and face a number of health challenges. For instance, in Russia, it is noted that 'Vepsians have poorer health compared to the non-indigenous people residing in the Barents Russia' (Emelyanova and Rautio 2016, 9). By comparison, mortality rates of the Sami in Finland are similar to the national mortality rate. On the other hand, mortality rates were lower in the 1980s for Sami in Finland, but rates have shifted over the last 30 years to become similar to the rest of the country and other Barents territories (Emelyanova and Rautio 2016, 9). More specifically, 'cancer rates were especially low, however they are now equal to the average values in Finland and Lapland' (Emelyanova and Rautio 2016, 9). According to Emelyanova and Rautio, 'this may be caused by changes in the habits and living environment of the Sami occurring in the period from 1970s up to nowadays, which are now similar to the majority of Finnish and western populations' (Emelyanova and Rautio 2016, 9). However, further analysis and data are needed in order to reach any substantial conclusions. Additionally, 'mortality due to accidents and violence is also significantly higher for the Sami than the national average. Non-fatal accidents and suicide attempts are also more common for Sami males' (Emelyanova and Rautio 2016; Soininen 2015). Finally, indigenous peoples in the Barents Region are also among peoples who face the highest number of suicides, especially in Nenets areas where the suicide rate for Nenets is significantly higher than the non-indigenous rate (Sumarokov et al. 2014) These differences have been partly explained by 'a lack of a "sense of indigenous belonging", lack of cultural identity and problems of resilience', being single or divorced, and having lower education (Sumarokov et al. 2014; Emelyanova and Rautio 2016, 10). The high levels of suicide in the Barents Region among indigenous peoples are also more broadly reflected in global trends that note 'indigenous people around the world have the highest suicide risk of any identifiable cultural (or ethnic) group' (Leenaars 2006).

In the Barents Region, certain climate factors affect the health and wellbeing of the Barents population differently than the national population. Seasonal statistics for morbidity show for instance 'that more deaths occur in the winter (cardiovascular, respiratory, circulatory, and skin diseases), and there are more traumas (frostbite and hypothermia) associated with cold weather' (Emelyanova and Rautio 2016, 10). In northern Sweden, it was also observed that 'a one-degree increase in temperature has led to a steep rise in the number of cases of non-lethal, acute myocardial infarction, and other heart dysfunctions' (Messner et al. 2002; Emelyanova and Rautio 2016, 8).

Thus, notwithstanding the marked differences between the Nordic and Russian parts of the region, the Barents Region is impacted by unique regional circumstances that influence the wellbeing of its population. The cold, humid and dark ambient environment in the Barents Region can make work and life more demanding for inhabitants, especially for individuals with disabilities or existing health issues. Long travel distances in the Barents Region outside urban areas also increase the costs of health care, disaster prevention and rescue services. In addition, the impact of environmental pollution and climate change has been the main focus of recent projects regarding health security challenges faced by the Arctic population. The Arctic Monitoring and Assessment Programme (AMAP 2014) and the EU-funded research project Arctic Health Risk (ArcRisk 2013) have monitored health security risks in relation to climate impacts (extreme temperatures, water scarcity, flooding, sea level rise, droughts, storms and ice and snow cover changes) and environmental contaminants. In the Barents Region, there have also been intensive efforts to enhance research and cooperation in order to address issues related to health and wellbeing in relation to the impact of the changing environment.

In order to strengthen health and wellbeing in the Barents Region, a working group on health and social issues has also been created under the auspices of the Barents cooperation. Established in 1996, the role of the Joint Working Group on Health and Related Social Issues (JWGHS) is to improve the public health and social wellbeing of peoples in the Barents Region. The JWGHS 'provides a political structure for health and social cooperation, in which representatives of the national and regional health authorities of the Barents countries meet twice a year to discuss relevant issues' (JWGHS 2017c). In its programme, the prioritised areas of intervention include the prevention and control of communicable diseases and non-communicable diseases, including work on the reduction of lifestyle-related risk factors, environmental factors and new emerging risks. Other priorities areas also concern the strengthening of health systems and social services with relevance for health in the Barents (JWGHS 2016, 2). These reflect the main challenges faced by the Barents population today and are highlighted in the following part of this assessment.

3 Assessment

The purpose of this assessment is to highlight several of the specific and distinct health security risks that face the Barents population. Specifically, the analysis allows focuses on the impact of different phenomena that affect the health and wellbeing of the Barents population such as the impact of the unique Barents climate and environment, industrial pollution and new environmental, social and economic changes.

3.1 Climate conditions and remoteness

The Arctic-Barents Region possesses a unique climate and a specific environment, where access to health care in remote areas can be difficult and affect the wellbeing of its population.

In the Barents Region, cold climate and darkness during the long wintertime is a challenge for the people living and working in the area. Cold ambient temperature can expose individuals to significant health risks. Cold sensations can be painful and cold exposure can worsen symptoms of some medical disorders (Parsons 2003; Rintamäki 2007a; Rintamäki 2007b). Body cooling under 35°C is associated with the range of physiological responses and dysfunctions (Parsons 2003). Working and surviving in demanding circumstances in cold, humid and dark ambient environments can require great effort, especially when a person has existing health problems.

In addition, cold weather and lack of light in wintertime can also have indirect effects on mood and cognition, which in turn effect the pineal and thyroid hormones that control psychological factors (Parsons 2003). Finally, accidental hypothermia, accidents and cold injuries also increase mortality in cold regions. This is particularly challenging for elderly people, whose thermoregulatory mechanisms may be weakened (Mallet 2002, 776). Finally, it can be noted that the cold ambient temperatures, wind, ice and snow conditions in the Barents add additional logistic challenges for health care authorities and rescue services to operate.

The sparsity and spread of populations in the Barents Region also has an impact on access to healthcare. In the northern parts of the Nordic countries and the Russian Barents Region, 'the organisation of the health service is affected by the fact that this is a very sparsely populated area with long distances' (BEAC 1998, 8). Consequentially, primary health care treatment in the Barents Region is predominantly centralised, where specialised treatment is only available through hospitals in concentrated urban areas (BEAC 1998, 8). This means that patients must sometimes travel long distance to receive necessary health care. Conversely, doctors may also travel long distances to visit their patients. For example, in the Yamalo-Nenets Autonomous region in Siberia, medical personnel sometimes travel to meet patients living in the tundra region (Pääkkölä 2013). In addition, emergency response times in remote areas are longer than in urban areas, and are exacerbated by harsh weather conditions. Delays in response time can occur when there are longer distances between hospitals in urban centres and emergencies in remote areas. Access to health care and emergency services in such remote areas depend directly on transportation capacities: helicopters, ambulances, snowmobiles, boats and ships.

On the other hand, there are solutions to mitigate the regional health impacts faced by the Barents population. First, the digitalisation of the health care sector, telehealth services and telemedicine can provide more equitable access and opportunities for all health care users. The development of portable equipment, measurement tools and facilities to support such digitalisation can also extend good quality e-health services to rural areas. However, training for emergency responders to use these new tools and methods is necessary in order to ensure their utility. In addition, while the development of new technologies, automations, the internet of things, 3D printing and other innovations are enticing, their impact on everyday life in the Barents is still relatively low. In this regard, it can be argued that more resources need to be allocated to develop future ICT-based services in relation to healthcare sectors.

Furthermore, cross-border cooperation among the Barents countries is an important means to solve challenges to healthcare in the region, especially in securing access to healthcare for communities in remote areas. Today, the establishment of the Joint Committee on Rescue Cooperation in the Barents Region

> makes it possible to provide assistance more efficiently, faster and at lower operating cost. Under the auspices of this cooperation, direct access to additional resources and specialised functions in neighbouring countries can be assured and preparedness and prevention measures can be co-ordinated.
>
> (BEAC 2017)

Yet, because of many new challenges such as climate change, it is also noted that Barents representatives should 'be aware and prepare for the need for enhanced cooperation in the Barents Region' (BEAC 2016). Among these challenges, the impacts of climate change and environmental pollution are threats that can directly affect the immediate and long-term health of the Barents population.

3.2 Environmental pollution – regional and transboundary contaminants

Regional and transboundary pollution represents other sources of health insecurity in the Barents Region. The Arctic, including the Barents, is contaminated by chemicals released as a result of both human activities and natural processes (e.g., AMAP 2011, Sundseth 2015, 3581). These chemicals reach the Arctic from regional and distant sources via the atmosphere and via northerly flowing rivers and ocean currents. Long-range transportation of the chemical contaminants as well as the presence of local contaminants cause a risk to the environment and affect human health in the Barents areas (see also Chapter 2.1). More specifically, environmental contaminants include a number of Persistent Organic Pollutants (POPs) (AMAP 2004) and mercury (AMAP 2011), which have been proven to affect human health. In general,

> because they can be transported by wind and water, most POPs generated in one place or country can and do affect people and wildlife far from where they are used and released. They persist for long periods of time in

> the environment and can accumulate and pass from one species to the next through the food chain.
>
> (Girard 2010, 404)

Concerning the impact of contaminants on human health, it can generally be stated that certain environmental pollutants can adversely affect the development of the immune system (AMAP 2015, 97). In the Russian Barents, industrial pollution has also been linked to an increase in mortality caused by respiratory problems, cardiovascular and circulatory diseases, and cancer (Nieminen et al. 2013; Norseth 1994, 106–107; Revich 1995). Other studies have shown that toxic agents and pollutants, such as organochlorine contaminants (OC), may have negative effects on human body and even increase obesity and diabetes (Lee et al. 2007; Longnecker et al. 2001; Longnecker and Daniels 2001; Kristiansen and Frøyland 2010; Son et al. 2010). The ArcRisk project (2013) has also studied the impacts of environmental contaminants and the complex relationships between sources, transport, bioaccumulation and exposure of contaminants on human health in the Arctic. Its conclusions indicate that human dietary exposure to harmful contaminants like PCBs, DDTs and HCHs has decreased lately due to the decrease of contaminants in environmental levels but also due to changes in dietary habits (AMAP 2014, 10–14). Nevertheless, global emissions of mercury as well as newly emerging contaminants still pose problems, especially for people living in the coastal communities (AMAP 2014, 10).

One of the worrying implications of transboundary pollution also concerns its impact on maternal and child health. Persistent toxic substances like mercury or like polychlorinated biphenyls and pesticides may threaten the health of the unborn and children (AMAP 2015). It has been demonstrated that mercury can affect the reproductive, immune and neurological systems and even the cardiovascular systems (AMAP 2011, 165). The impact of pollutants on unborn and newborn health occurs through the passing of Persistent Organic Pollutants (POPs) in the food chains and breast milk. Breastfeeding is highly recommended as primary food for infants (WHO, n.d.). However, if the human milk contaminant burden increases alongside the presence of Arctic pollutants and toxic chemicals in food chains in fish or mammals, the practice of breastfeeding may become a more complex public health issue. In particular, this could be a major concern among indigenous peoples living in tundra environment in Russia, far from the other modern food supplies (see also Chapter 2.4). In this regard, further research and general awareness for the complexity of the impacts of increasing contaminant burdens in the Arctic and Barents Region are needed.

Finally, apart from its toxicological risks, the perception and knowledge of potential risks associated with contaminants in the environment can also adversely 'turn people away from traditional/local foods, even when they represent the most healthy food choice' (AMAP 2011, 164). As noted in the 2011 AMAP report, 'this can lead among other things, to a degradation of cultural identity, economic stress arising from the high cost of healthy store-bought foods, and potential nutritional deficits arising from the consumption of poor quality store-bought foods'. Thus, pollution in the Barents Region can have a more broad and indirect impact on the health and wellbeing of its population, especially in regards to communities with specific traditional foods such as indigenous communities.

Ultimately, continuous research is needed to understand how environmental pollution affects the health of peoples living in the North, especially in the Barents-Arctic where more data should be collected to understand the current impact of pollution on the human health of the population. In any event, there is a need to reduce contaminant levels and pollution through local, national, regional and international actions and initiatives (AMAP 2011, 167–169).

3.3 Environmental, economic and social changes

Ecosystem disturbance and changing interactions between wildlife and humans can lead to the spread of new pathogens. Today, worldwide, there is an apparent increase in many infectious diseases, including newly circulating ones. This reflects the combined impacts of rapid demographic, environmental, social, technological and other lifestyle changes that affect both physiological and mental wellbeing.

3.3.1 Antibiotic resistant microbes and virus

The presence of, and risk posed by, antibiotic resistant microbes in the Arctic are increasing (European Center for Disease Prevention and Control 2017). In the Barents Region, biological factors such as mutations in bacteria and the development of the super bacteria have resulted in the presence of multi-drug resistant tuberculosis bacteria in Russia and in neighbouring parts of Finland (Casali et al. 2014, JWGHS 2012). Thus, according to the Joint Working Group on Health and Related Social Issues (JWGHS), 'the epidemiological situation regarding tuberculosis in the Barents Region is rather controversial … and its threat is increasing …' (JWGHS 2012, 4).

Specifically, tuberculosis (TB) incidence in the Nordic countries is among the lowest in the world, which, however contrasts with Russia, which is classified as one of the high-priority countries for TB by WHO. In Finland, the incidence of tuberculosis is low, but the risk of TB resistance is increasing, especially among foreign nationals. In 2015, all children who contracted TB were of foreign origin, and other adult patients were reported to be foreign in 105 cases (39%), up by one-fifth year-to-year (Jaakola et al. 2015, 44) According to Jaakola, 'the growing numbers of asylum seekers were probably one of the contributing factors to this trend' (Jaakola et al. 2015, 5). As a consequence, Finland has started to carefully check the health of newcomers as a preventative measure. Although it is unclear how this trend affects more specifically the Barents part of Finland, it is certainly an issue that affects this part of the country as well.

Unlike Northern Europe, where TB is mostly registered among immigrants and elderly people, 'the epidemic process in Russia penetrates into various population layers, becoming a problem not only of vulnerable groups, but also of socially advantaged ones' (JWGHS 2012, 4). Due to this situation, control over TB has been restored in the country, which has allowed a stabilisation in TB incidence, prevalence and mortality in the last years, with a tendency to improve the epidemic situation (JWGHS 2012, 4). Nonetheless, the JWGHS has suggested taking measures to prevent the spread of tuberculosis and HIV&TB co-infection in the Barents Region through intensive international collaboration and to develop guidelines, standards and infectious control programs for the Russian part of the Barents Region (JWGHS 2017a). In addition, and among other measures, the working group has also been advised 'to improve the implementation of the existing plans for infection control', 'to introduce earlier and more accurate diagnostics of TB and MDR TB' and 'to improve capacities of facilities and institutions' (JWGHS 2017b).

Another area of significant concern in the Barents Region is the Human Immunodeficiency Virus (HIV). In 2013, 6,472 new cases of HIV infection were reported in eleven territories of the Northwestern Federal District of Russia. Although the number of case had decreased from 2.1% since 2012, an increase in the number of new HIV cases was reported in four areas of the North-western Federal District: Arkhangelsk (by 15.2%), Vologda (by 12.1%), Pskov (by 7.4%) and Murmansk (by 2.8%) regions (JWGHS 2015, 9). 'Since the beginning of the epidemic in 1989, the number of officially registered HIV cases in the NWFD total 102,623. During the same period, 17 283 people have died' (JWGHS 2015, 15). As the HIV epidemic continues to spread in the Barents Region of Russia under some notable trends (JWGHS 2015, 15), a better action program is urgently needed to tackle the issue. Yet, in its 2015 report, the JWGHS indicates that 'a part of the population still has improper awareness about HIV/AIDS' and that 'the coverage of the harm reduction programs remains modest'. In addition, there is a lack of funding for awareness and harm reduction programmes, which constitutes a major hurdle in addressing the epidemic (2015, 16).

By comparison, the HIV infection rate reported in all three Nordic countries is relatively stable, and the total number of diagnoses by doctors in Norway is even in decline (JWGHS 2015). In all cases reported in the Nordic countries, foreigners and newly arrived migrants represent more than half of the diagnosed patients (JWGHS 2015, 13). In this regard, the JWGHS has underlined 'the importance of developing efforts and methods to reach migrants, for HIV preventive work, with information and to offer counselling and testing' (JWGHS 2015, 15). Overall, prevention programmes and policy support are needed across the Barents to address the spread of infectious diseases such as HIV.

3.3.2 Zoonotic, new and hidden diseases

The apparition of new species in the Arctic, such as new insects also raises health concerns. New zoonotic diseases may emerge in the Barents Region in the future (Revich et al. 2012). Many blood-sucking insects can act as vectors for microbes, and transmit diseases to humans and animals via bites such as mosquitoes and flies. Rising tick populations in the Barents are an on-going concern (Evengård and McMichael 2011; Sormunen et al. 2016; Jääskeläinen et al. 2016). In Russian Karelia it has been reported in 2012 that the incidence of Tick Borne Encephalitis disease remains high and the incidence in the republic is two to three time higher than the incidence for the country as a whole (EpiNorth 2016). Tularemia, another zoonotic disease caused by the bacteria *Francisella tularensis,* is also an issue of concern in the northern part of Finland and the number of human cases of tularemia increased significantly in Sweden between 1984–2012 (Rossow 2015; Furberg 2016).

Additionally, temperature changes around the globe are pushing human pathogens towards the Arctic, raising many new risks for peoples. Previous studies suggest that the climate change will affect the relationship between regional climates and allergens, especially on aeroallergens (pollen and mould spores). Climate change will have impacts on four categories of allergens – respiratory, contact allergens, plant food allergens, insect allergens (Beggs 2014, 105–113). Increasing temperatures will change the spatial patterns of parasites including house dust mites (a common allergen in the home) and birch pollen (Frei and Gassner 2008). Increased rainfalls as well as an increase in relative humidity indoors might cause difficulties in managing excess water flows in buildings and contribute to the risk of mould growth, which could impact respiratory health. Studies regarding the effects of climate change on respiratory health have been restricted to the last two decades, (Beggs 2014, 105) and there is a great need for this research to continue globally as well as in the context of the Barents Region.

Climate change is also likely to accelerate the spread of new pathogens including tropical diseases, which has led researchers to view the longstanding relationships between climate and disease with new urgency (Institute of Medicine 2008). Yet, little is known about the impact of climate change and the risk and distribution of infectious diseases in Arctic regions. According to Parkinson and Evengård (2009), damage to sanitation infrastructures and water sources caused by melting permafrost, storm surge or flooding can adversely affect water quantity and quality, and result in increased rates of respiratory and skin infections as well as diarrheal diseases caused by bacterial, viral and parasitic agents. More particularly, climate change can spread illnesses such as malaria, as rising temperatures push disease-carrying mosquitoes into new places. The tropical disease malaria, a vector-borne disease caused by the malaria parasite (*Vivax falciparum*), was a common endemic disease in Finland in the 18th and 19th centuries (Hulden and Hulden 2009). However, with global warming temperature, it is possible that the virus will resurface.

Finally, hidden diseases such as the bacteria *Bacillus anthracis* may also reappear. In fact, *Bacillus anthracis* resurfaced in 2016 and caused an epidemic amongst reindeer in the Yamalo-Nenets region of Siberia in Russia (BBC News 2016). According to many scientists, the resurfacing of hidden diseases may increase as icy environments melt and ancient microbes emerge from the ice and mix their genes with the modern microbial genomes (Katz 2012; Revich et al. 2012).

Thus, the (re)emergence of infectious diseases in the Barents Region depends on many factors. Suitable conditions for disease transmission may change with the environment and across time. However, changes in the spatial patterns of diseases remain hard to estimate, although scientists are currently studying the spreading patterns of many potential and infectious microbes that pose health risks both globally and within the Barents Region (e.g. Revich et al. 2012; Bouma and Pascual 2014).

3.3.3 Risk and disasters

Natural disasters in the Barents are rare. Nevertheless, climate change increases the uncertainty for future risks. Quick changes in weather conditions and extreme weather events due to the climate change may create new risks for the inhabitants in the Siberian tundra and in Fennoscandia (IPCC 2014). The melting of ice and permafrost will increase the difficulty in forecasting future weather patterns, and may result in increasing incidences of quickly changing cold ambient temperatures, wind, ice and snow conditions in the Barents.

The uncertainty of climate-related changes to regional weather may make it more difficult for authorities responsible for regional health care and rescue services to be prepared in the future. In this respect, strengthening cross-border cooperation among the Barents countries is important. The Arctic Council is already promoting search and rescue cooperation between to improve rescue services in the region. In 2011, it adopted the Search and Rescue Agreement (formally the Agreement on Cooperation on Aeronautical and Maritime Search and Rescue in the Arctic) to coordinate international search and rescue coverage and response in the Arctic. Although the Barents Region is included in the Search and Rescue Agreement, regionally the Barents also organises rescue exercises biannually. The Barents Rescue is a collaborative exercise organised under the auspices of the Barents Joint Committee on Rescue Cooperation. It seeks educates people working in rescue services, as well as local actors in business enterprises and municipalities, and prepares them for cross-border accidents and emergencies. Both the activities of the Arctic Council and Barents cooperation are essential to prevent accidents in the Arctic, as potential emergencies may increase with environmental changes or developments in the maritime and industrial sectors.

While the impacts of climate change are difficult to assess comprehensively, extreme weather and environmental changes can have serious impacts on the health and wellbeing of people living in the Barents, especially because many of the residents' livelihoods are connected to nature. Ultimately, further research to understand new phenomena and avenues to address potential changes collaboratively are needed.

3.3.4 Physiological and mental wellbeing

Alongside environmental changes, societal changes linked with the impact of the market economy and globalisation impacts the human security of peoples living in the Barents, including their health.

In particular, significant changes in diet through the decrease in traditional food uses and increase in the use of imported food are both factors in increasing food and health insecurities. Various nutrition related problems have been observed in the Arctic, and the effects of changing diets on health are being investigated through scientific research in many countries worldwide (AMAP 2011; Sharma et al. 2010). For instance, the incidence of obesity, increased body weight and chronic diseases including diabetes and cardiovascular diseases are growing in the circumpolar Arctic. Obesity is a major contributor to the global burden of chronic diseases and disabilities and is closely connected to serious social and psychological disorders affecting all ages and socioeconomic groups (Hansen et al. 2014). Several factors explain the increased rate of obesity including biological, genetic and epigenetic factors and cultural adaptations. Higher indoor temperatures (thermic-neutral temperatures), a lack of physical activity also decrease natural energy expenditure and can cause a higher risk of obesity. The societal shift towards more readily available high energy density foods can also contribute to increased rates of obesity. Furthermore, the shift from traditional to western foods have been linked to increased rates of obesity in Nenets communities (Hopping et al. 2010; Petrenya et al. 2014).

Furthermore, psychological health also clearly affects mental wellbeing. The resilience of the people living in the Barents, and the effect of such rapid changes on the long-term wellbeing of peoples and the human mind is still unknown. An increased sense of grief and the loss of control can cause problems in mental health and increase mental illnesses (Evengård and McMichael 2011). Suicide rates in the Northern areas are already relatively high, especially among indigenous communities, (Evengård and McMichael 2011; Bjerregaard and Larsen 2015; Young et al. 2015) and mental health services are not always easily accessible. Limited medical services in indigenous languages are also an obstacle for the use of psychiatric health care services in some areas. Finally, violence and harassment are also a growing problem in the Arctic and Barents Region (See also Chapter 2.7 on Personal Security).

4 Conclusion

Health security in the Barents Region is undoubtedly evolving in response to increasing societal and environmental changes. As discussed in the previous sections, globalisation, environmental changes and demographic shifts are keys factors affecting the health status of the Barents population. Migration, population aging, urbanisation and changing socio-economic conditions may also accelerate the emergence of the health problems induced by the changing climate and increasing extreme weather events (IPCC 2014).

Whereas the health concerns faced by peoples living in the Russian Barents may differ in scale and nature from those in the Nordic countries, the Barents population still share similar challenges. Addressing these challenges requires national and regional cooperation both to study the development of present and future health risks and to raise awareness among the population. Increased cooperation, digitalisation and new technological solutions to guarantee secured and efficient health services for all in the Barents Region are needed. In addition, holistic understandings and approaches to health security are required, in order to account for psychological, physiological and psychosocial wellbeing (Commission on Human Security 2003, 95). Finally, the lack of comprehensive data and studies concerning the present and future health issues concerning the Barents Region more specifically illuminates the need to enhance research in this field in order to improve access and adequate health care for all in the region.

References

AMAP (Arctic Monitoring and Assessment Programme). 2004. *AMAP Assessment 2002: Persistent Organic Pollutants in the Arctic*. Oslo.

AMAP. 2011. *AMAP Assessment 2011: Mercury in the Arctic*. Oslo.

AMAP. 2014. *ArcRisk. Arctic Health Risks: Impacts on Health in the Arctic and Europe Owing to Climate-induced Changes in Contaminant cycling. Results Overview*. Oslo.

AMAP. 2015. *AMAP Assessment 2015: Human Health in the Arctic*. Oslo.

ArcRisk. 2013. *ArcRisk: Results Overview*. www.arcrisk.eu/

BBC News2016. 2 August. www.bbc.com/news/world-europe-36951542

BEAC. 1998. *Health Statistic Indicators for the Barents Euro-Arctic Region*. https://norden.diva-portal.org/smash/get/diva2:968672/FULLTEXT01.pdf

BEAC. 2016. *Barents Rescue 2015 Exercise. Evaluation Report*. www.pelastustoimi.fi/download/68586_BR_15_Evaluation.pdf?942dcc010f5bd488

BEAC. 2017. *Joint Committee on Rescue Cooperation in the Barents Region*. http://www.barentscooperation.org/en/Working-Groups/BEAC-Working-Groups/Rescue-Cooperation.

Beggs, P.J. 2014. Impacts of climate change on allergens and allergic diseases: knowledge and highlights from two decades of research. In *Climate Change and Global Health*, edited by C.D. Butler. Wallingford: CABI.

Bjerregaard, P. and Larsen, C.V.L. 2015. Time trend by region of suicides and suicidal thoughts among Greenland Inuit. *International Journal of Circumpolar Health* 74: 26053.

Bouma, M. J. and Pascual, M. 2014Global warming and malaria in tropical highlands – an estimation of Ethiopia's 'unmitigated' annual malaria burden in the 21st century. In *Climate Change and Global Health*, edited by C.D. Butler. Wallingford: CABI.

Casali, N., Nikolayevskyy, V., Balabanova, Y., Harris, S.R., Ignatyeva, O., Kontsevaya, I., Corander, J., Bryant, J., Parkhill, J., Nejentsev, S., Horstmann, R.D., Brown, T. and Drobniewski, F. 2014. Evolution and transmission of drug-resistant tuberculosis in a Russian population. *Nature Genetics* 46(3): 279–286.

Commission on Human Security. 2003. *Human Security Now*. New York: UN Trust Fund for Human Security.

Elbe, S. 2010. *Security and Global Health*. Cambridge: Polity Press.

Emelyanova, A. and Rautio, A. 2016. Population iversification in demographics, health, and living environments: the Barents Region in review. In *NGP Yearbook 2016: Geographies of Well-being in the North*. Edited by T. Lankila and K. Tervo-Kankare. Oulu, Finland: Nordia Geographical Publications.

EpiNorth. 2006. *Epidemiological Trends of Communicable Diseases in the Baltic Sea and Barents Regions, 2000*. www.epinorth.org/eway/default.aspx?pid=230&trg=Area_5268&MainArea_5260=5263:0:15,2946:1:0:0:::0:0&Area_5263=5268:44984::1:5264:1:::0:0&Area_5268=5273:45469::1:5266:3:::0:0

European Centre for Disease Prevention and Control. 2017. *Antimicrobial Resistance Surveillance in Europe 2015*. Stockholm: Annual Report of the European Antimicrobial Resistance Surveillance Network (EARS-Net).

Evengård, B. and McMichael, A. 2011. Vulnerable populations in the Arctic. *Global Health Action* 4: 3–5.

Field, C.B., Barros, V.R., Mastrandrea, M.D., Mach, K.J., Abdrabo, M.A.K., Adger, W.N., Anokhin, Y.A.et al. 2014. Summary for policymakers. In *Climate Change 2014: Impacts, Adaptation, and Vulnerability. Part A: Global and Sectoral Aspects. Contribution of Working Group II to the Fifth Assessment Report of the Intergovernmental Panel on Climate Change*, edited by C.B. Field, V.R. Barros, D.J. Dokken, K.J. Mach, M.D. Mastrandrea. IPCC.

Frei, T. and Gassner, E. 2008. Climate change and its impact on birch pollen quantities and the start of the pollen season an example from Switzerland for the period 1969–2006. *International Journal of Biometeorology* 52(7): 667–674.

Furberg, M. 2016. *Towards the Limits – Climate Change Aspects of Life and Health in Northern Sweden: studies of tularemia and regional experiences of changes in the environment*. Doctoral dissertation. Umeå: Umeå University Press.

Girard, J.E. 2010. *Principles of Environmental Chemistry*. 2nd edition. Sudbury: Jones and Bartlett Publishers.

Hansen, J.C., Gilman, A.P. and Odland, J.O. 2014. *The Seven Pillars of Obesity: Development of a Holistic Understanding of a Complex Problem. AMAP Technical Report No. 8*. Oslo: Arctic Monitoring and Assessment Programme (AMAP).

Hopping, B.N., Erber, E., Beck, L., De Roose, E. and Sharma, S. 2010. Inuvialuit adults in the Canadian Arctic have a high body mass index and self-reported physical activity. *J Hum Nutr Diet* 23 (Suppl. 1): 115–119. doi:10.1111/j.1365–277X.2010.01103.

Hulden, L. and Hulden, L. 2009. The decline of malaria in Finland – the impact of the vector and social variables. *Malaria Journal* 8(94).

Institute of Medicine (US). 2008. *Forum on Microbial Threats, Summary and Assessment. National Academies Press (US)*. www.ncbi.nlm.nih.gov/books/NBK45737/.

IPCC. 2014. *Climate Change 2014 Synthesis Report Summary for Policymakers*. IPCC.

Jaakola, S., Lyytikäinen, O., Rimhanen-Finne, R., Salmenlinna, S., Pirhonen, J., Savolainen-Kopra, C., Liitsola, K.*et al.*2015. *Infectious Diseases in Finland 2015. Report 15/2016*. Helsinki: National Institute for Health and Welfare (THL).

Jääskeläinen, A. et al. 2016. Siberian subtype tick-borne encephalitis virus in Ixodes ricinus in a newly emerged focus, Finland. *Ticks and Tick-borne Diseases* 7(1): 216–223.

JWGHS. 2012. *The Barents Tuberculosis Program*. www.barentsinfo.fi/beac/docs/Barents_Tuberculosis_Program_2013_ENG.pdf

JWGHS. 2015. *The Barents HIV/AIDS Programme 2015–2019*. www.barentsinfo.fi/beac/docs/Barents_HIV-AIDS_programme_2015-2019_ENG.pdf

JWGHS. 2016. *5th Framework Programme for Cooperation on Health and Related Social Issues in the Barents Euro-Arctic Region 2016–2019*. BEAC.

JWGHS. 2017a. *Barents HIV/AIDS Programme*. www.barentscooperation.org/en/Working-Groups/Joint-Working-Groups/Health-and-Social-Issues/Barents-HIVAIDS-Programme

JWGHS. 2017b. *Barents Tuberculosis Programme*. www.barentscooperation.org/en/Working-Groups/Joint-Working-Groups/Health-and-Social-Issues/Barents-Tuberculosis-Programme

JWGHS. 2017c. *Health and Social Issues*. www.barentscooperation.org/en/Working-Groups/Joint-Working-Groups/Health-and-Social-Issues

JWGHS Statistic Working Group. 1998. *Health Statistic Indicators for the Barents Euro-Arctic Region*. Copenhagen: Nordic Medico-Statistical Committee (NOMESCO). https://norden.diva-portal.org/smash/get/diva2:968672/FULLTEXT01.pdf

Katz, C. 2012. Melting glaciers liberate ancient microbes. The release of life-forms in cold storage for eons raises new concerns about the impacts of climate change. *Daily Climate*, 18 April.

Kristiansen, K. and Frøyland, L. 2010. Persistent organic pollutant exposure leads to insulin resistance syndrome. *Environmental Health Perspectives* 118(4): 465–471.

Lee, D.H., Lee, I.K., Jin, S.H., Steffes, M. and Jacobs, D.R.Jr. 2007. Extended analyses of the association between serum concentrations of persistent organic pollutants and diabetes. *Diabetes Care* 30(6): 1596–1598.

Leenaars, A. 2006. Suicide among indigenous peoples: introduction and call to action. *Archives of Suicide Research: Official Journal of the International Academy for Suicide Research* 10(2): 103–115.

Longnecker, M.P. and Daniels, J.L. 2001. Environmental contaminants as etiologic factors for diabetes. *Environmental Health Perspectives*, 109 (suppl 6): 871–876.

Longnecker, M.P., Klebanoff, M.A., Brock, J.W. and Zhou, H. 2001. Polychlorinated biphenyl serum levels in pregnant subjects with diabetes. *Diabetes Care* 24(6): 1099–1101.

Mallet, M.L. 2002. Pathophysiology of accidental hypothermia. *An International Journal of Medicine* 95(12): 776.

Messner, S. F., Raffalovich, L. E. and Shrock, P. 2002. reassessing the cross-national relationship between inequality and homicide rates: implications of data quality control in the measurement of income distribution. *Journal of Quantitative Criminology* 18: 377–395.

Nieminen, P., Panychev, D., Lyalyushkin, S., Komarov, G., Nikanov, A., Borisenko, M., Kinnula, V.T. and Toljamo, T. 2013. Environmental exposure as an independent risk factor of chronic bronchitis in northwest Russia. *International Journal of Circumpolar Health* 72: 19742.

Norseth, T. 1994. Environmental pollution around nickel smelters in the Kola Peninsula (Russia). *Science of the Total Environment* 148(2–3): 103–108.

Pääkkölä, S. 2013. Personal discussion with MD Andrei Lobanov in Scientific Research Centre of the Arctic. Russia. Yamalo-Nenets Autonomous Okrug, Nadym city (65° 31′N 72° 31′E) 27 November.

Parkinson, A.J. and Evengård, B. 2009. Climate change, its impact on human health in the Arctic and the public health response to threats of emerging infectious diseases. *Global Health Action* 2.

Parsons, K. 2003. *Human thermal environments: The Effects of Hot, Moderate and Cold Environments on Human Health, Comfort and Performance.* London: Taylor & Francis CRC Press.

Petrenya, N., Brustad, M., Dobrodeeva, L., Bichkaeva, F., Lutfalieva, G., Cooper, M. and Odland, J.Ø. 2014. Obesity and obesity-associated cardiometabolic risk factors in indigenous Nenets women from the rural Nenets Autonomous Area and Russian women from Arkhangelsk city. *International Journal of Circumpolar Health* 73.

Revich, B.A. 1995. Public health and ambient air pollution in Arctic and subarctic cities of Russia. *Science of the Total Environment* 160/161: 585–592.

Revich, B., Tokarevich, N. and Parkinson, A.J. 2012. Climate change and zoonotic infections in the Russian Arctic. Review article. *International Journal of Circumpolar Health* 71.

Rintamäki, H. 2007a. Human responses to cold. *Alaska Medicine* 49(2): 29–31.

Rintamäki, H. 2007b. Performance and energy expenditure in cold environments. *Alaska Medicine* 49(suppl 2): 245–246.

Rossow, H. 2015. Epidemiology of tularemia in Finland. Academic dissertation. Helsinki: Faculty of Veterinary Medicine, University of Helsinki.

Sharma, S., Gittelsohn, J., Rosol, R. and Beck, L. 2010. Addressing the public health burden caused by the nutrition transition through the Healthy Foods North nutrition and lifestyle intervention programme. *J Hum Nutr Diet* 23(Suppl. 1): 120–1127. doi:10.1111/j.1365–277X.2010.01107.

Soininen, L. 2015. The health of the Finnish Sami in light of mortality and cancern pattern. Academic dissertation. Helsinki: University of Helsinki.

Son, H.K., Kim, S.A., Kang, J.H., Chang, Y.S., Park, S.K., Lee, S.K., Jacobs, D.R.Jr and Lee, D.H. 2010. Strong associations between low-dose organochlorine pesticides and type 2 diabetes in Korea. *Environment International* 36(5): 410–414.

Sormunen, J.J., Klemola, T., Vesterinen, E.J., Vuorinen, I., Hytönen, J., Hänninen, J., Ruohomäki, K., Sääksjärvi, I.E., Tonteri, E. and Penttinen, R. 2014. Assessing the abundance, seasonal questing activity, and Borrelia and tick-borne encephalitis virus (TBEV) prevalence of Ixodes ricinus ticks in a Lyme borreliosis endemic area in Southwest Finland. *Ticks and Tick-borne Diseases* 7(1): 208–215.

Sumarokov, Y., Brenn, T.Kudryavtsev, A.V. and Nilssen, O. 2014. Suicides in the indigenous and non-indigenous populations in the Nenets Autonomous Okrug, Northwestern Russia, and associated socio-demographic characteristics. *International Journal of Circumpolar Health* 73.

Sundseth, K., Pacyna, J.M., Banel, A., Pacyna, E.G. and Rautio, A. 2015. Climate change impacts on environmental and human exposure to mercury in the Arctic. *International Journal of Environmental Research and Public Health* 12(4): 3581.

UN General Assembly. 2016. *High-level Meeting on Antimicrobial Resistance.* www.un.org/pga/71/event-latest/high-level-meeting-on-antimicrobial-resistance/

United Nations Development Program. 1994. *Human Development Report 1994.* New York: UN.

United Nations. 2007. *Indicators of Sustainable Development: Guidelines and Methodologies.* New York: UN.

WHO (n.d.). *Exclusive Breastfeeding*. WHO. www.who.int/nutrition/topics/exclusive_breastfeeding/en/.

Young, R.S.*et al.*2015. The frequent evolutionary birth and death of functional promoters in mouse and human. *Genome Research* 25: 1546–1557.

2.4 Food security in the Barents Region

Shaun Cormier and Dele Raheem

This chapter addresses food (in)security in the Barents Region. Food insecurity is interlinked with political, economic, health, environmental, personal, and community insecurities as indicators of human security.

1 Definition

The United Nations Food and Agricultural Organisation's current definition describes that, 'food security exists when all people at all times have physical, social and economic access to sufficient, safe and nutritious food to meet their dietary needs and food preferences for an active and healthy life' (FAO 1996). This definition takes into account the 1994 UNDP Human Development report that promoted food security as one of the threats to human security (UNDP 1994). Food security can be better understood through four main dimensions: availability, accessibility, utilisation and food systems stability. The availability of food is determined by the physical quantities of food that are produced, stored, processed, distributed and exchanged (FAO 2008, 20). Accessibility, according to the Food and Agriculture Organisation (FAO), is a measure of the ability to secure entitlements, which are those set of resources (legal, political, economic and social) that an individual requires to obtain access to food (FAO 2008, 20). Food utilisation refers to the appropriate nutritional content of the food and ability of the body to use it effectively (Burke and Lobell 2010: 14); in other words the safety and social value of food. Food systems stability refers to the overall balance of food supply and is determined by the temporal availability of and access to food (FAO 2008, 21). Food systems instability results when one or more of the four components of food security is uncertain and insecure.

In addition, food security contains terms that describe the values attached to foods such as local, traditional and store-bought. Local food, as defined by Duhaime and Bernard (2002, 228), refers to any food produced or harvested in the environment of a community, regardless of whether it was produced for subsistence or commercial purposes. Bertozzi (1998) described traditional foods as a representation of a group; it belongs in a defined space and is part of a culture, which implies the cooperation of the individuals operating in that territory (Jordana 2000, 147). Therefore, a traditional product or food is inherently linked to a territory and must be part of a set of traditions, which will necessarily ensure its continuity over time (Jordana 2000, 147). On the other hand, store food or store-bought food are frequently used terms designated to those foods that are imported to the Arctic-Barents Region (Duhaime and Bernard 2002, 228).

In the context of the Barents Region, it is important to consider that traditional food is a vital component of food security, especially for indigenous populations, for whom traditional foods are closely linked to their cultural identity and community security. For instance, oral tradition and rituals are associated with traditional hunting, fishing and gathering of these foods. In this way, the existing definition of food security is insufficient in assessing indigenous relationships to food, as it posits that food security relies entirely on the assessment of monetary access to foods purchased from markets, but the value of traditional foods for indigenous peoples exists beyond purely monetary assessments (Egeland and Harrison 2013, 17). Therefore, a human security approach to address threats undermining the resilience of diverse communities in relation to food security was introduced, as human security is people-centred, comprehensive, context-specific and prevention-oriented (FAO 2016, 4). Furthermore, this approach supports food sovereignty and the ability for locals to take more control of food systems in the Barents Region.

2 Contextualisation

An assessment of food security in the Arctic-Barents context should take note of the unique natural environment of the region. As discussed in previous chapters, the Barents Region has more forest, fish, minerals, oil and gas than any other region in Europe. This diversity and abundance of resources explains the existence of conflicting priorities between traditional food and human activities in the region.

Human activities such as fishing, hunting, agriculture, gathering of non-wood forest products and reindeer herding are common practices that are also important to the region's food security; such activities empower people to utilise local food crops, natural plants, games and fish. These practices encourage communal sharing and have historically not been quantified their monetary values. Indigenous peoples have practised these activities since the Middle Ages and non-indigenous peoples who eventually migrated to the region adopted similar practices from their interactions with indigenous communities (BEAC 2016).

A substantial seasonal variability has been described in the diet of Sámi indigenous peoples. Sámi communities in the 1930s–1950s mainly consumed meat during autumn and winter, but dried meat was also eaten in summer, particularly by men during work-related migration such as herding, hunting and rafting (Nilsson 2012, 40). Wild berries such as *Rubus chamaemorus* (cloudberry), *Vaccinium uliginosum* (bilberry), *Vaccinium vitis-idaea* (lingonberry), *Rubus idaeus* (raspberry), *Vaccinium oxycoccus* (cranberry), etc. are widely gathered, and are eaten both fresh or kept preserved during abundant seasons (Nilsson 2012, 40). Family gardens provide fresh potatoes, strawberries, blackcurrants, onions and root crops. Mushrooms and herbs picked from the forest are often part of the traditional meals, but on a much smaller scale. Reindeer meat is the most prevalent and important meat produced in the Barents Region; other native animals such as moose, rabbits, fowl and waterfowl are not as prevalent in meals because of the low number of wild game to hunt (Muller-Wille 2008, 347).

As societies in the Barents Region become increasingly globalised, food products are now constantly transported over long distances to remote communities in the region, resulting in rising consequences for food safety and security.

Globally, food production and agriculture accounts for 70% of all human water use (UN-Water 2015). Water and sanitation services are crucial in food processing, but are not well documented in the Arctic-Barents Region (Hennessy and Bressler 2016, 4). The safety and quality of food begins with the growing conditions of ingredients, including water. The quality of water provided by different municipalities in the Barents differ greatly between the Nordic countries and Russia, as the quality of water is not monitored with the same regularity across the Barents Region. Dudarev et al. (2013) called for an international collaboration in order to address water security and infectious disease issues, as they will have impacts on food security. Therefore, it is important to emphasise the relationship between water and food security in the region.

Traditional foods provide adequate nutrients for people in the Barents Region; however, these foods can also be a source of pollutants. In order to ensure that communities can adequately respond to ecological, environmental and social threats, an assessment of the complex role of traditional foods in the food security of the Barents Region is necessary

3 Assessment

Food security is threatened by climate change, human activity and globalisation in all four of its dimensions (FAO 2008, 18). These threats are complex and often interrelated, and a human security approach requires a holistic assessment. In the following sections, we briefly highlight how these threats play a role in the food security of the Barents Region.

3.1 Climate change

The rising variance of temperatures and weather patterns associated with climate change make it difficult to predict the availability, accessibility or stability of foods at a given time. Seasonal shifts affect the availability of food in the Arctic-Barents. Traditional foods from the soil, such as berries, shrubs and vegetables, are dependent on weather, and subject to alteration because of abnormal seasonal changes. Hunting game becomes more difficult; migration patterns adjust to irregular seasonal changes making it difficult to track reindeer, bowhead whale, fish and other bird species (FAO 2008). As the availability of these traditional foods become increasingly unpredictable, the dependency of communities on store-bought food increases in the Arctic-Barents Region (Duerden 2004, 207).

The environment is a crucial resource for obtaining access to food in the Arctic, and climate change could pose challenges to food accessibility. Attributed to changing weather patterns, the frequent thawing and freezing of snow in the Arctic has, for instance, hampered the reindeers' accessibility to food under the snow. Such outcomes of less access to pastures were documented in cases where autumn and winter rain-on-snow (ROS) events have resulted in ice-encrusted pastures and mass starvation of semi-domesticated reindeer (*Rangifer tarandus*) and have increased in frequency and intensity across the northwest Russian Arctic (Forbes et al. 2016, 1). This could be especially important for reindeer herders, who migrate with their reindeer during the summer and winter months to find appropriate feeding grounds for the herd. Therefore, in Russia the herds are often monitored closely by the herders to adapt quickly to the outcomes of climate change (Rees et al. 2008).

Apart from reindeer meat, the Sámi, Veps and Nenets in the Arctic-Barents Region rely on other traditional foods such as grouse, elk, fresh and saltwater fish, and berries; their access to such food is also changing with climate patterns (Berg 2014, 32). To this degree, climate change threatens the accessibility of traditional food for all indigenous peoples across the Barents Region.

Food utilisation is related to the nutritional content that the body needs to survive; traditional foods in the Barents Region are known to be high in protein, vitamins, oils, natural fats and other essential nutrients. Lack of traditional food due to limited accessibility and availability could promote further dependency on store-bought foods with a high intake of sugar, salt and chemical preservatives that can lead to further health issues such cardiovascular diseases, obesity, diabetes and dental cavities (van Oostdam et al. 2003). There is also growing evidence to suggest that warming temperatures in the Arctic can threaten the safety of food, as pathogens tend to thrive in this atmosphere (Burke and Lobell 2010, 28). This has implications for food safety throughout the whole food systems process, from hunting to storage, cooking and ingestion. Meat processing of game in the Arctic-Barents is not in a controlled environment with packaging and storage facilities, so the risk of pathogens and bacteria in food can be significantly higher (Parkinson and Evengård 2009, 2). An increase in the presence of bacteria in caught game could also occur if hunters have to account for additional travel to locate and transport food that is further away. Furthermore, many of the microbes to which marine mammals are susceptible are of concern to human health as they can be transferred between animals and humans (Burek et al. 2008, 132). In addition, the possibility of invasive species from the south migrating north due to warming temperatures can lead to biodiversity loss in the region which will have adverse effect on the food supply chain (Lassuy and Lewis 2013, 77).

Climate change can influence the overall stability of the local food system; today's globalised economy is highly sophisticated and thus far manages to supply the Barents Region with accessible and available food. However, this could change as the food system becomes more costly from the globalised impacts of climate change. Already, people in the Barents Region face the high cost of imported foods as they move away from traditional foods to a growing market-based economy. These changes have an impact on human health, livelihood assets, food production and distribution channels, as well as changing purchasing power and market flows (FAO 2008: iii). Furthermore, if these challenges persist, crises in food supply are expected to increase in intensity and duration, due to temporal fluctuations in food resources (Paci et al. 2004, 3). Chris Paci et al. (2004, 1) described climate change as a real and significant threat to food security of peoples living in the Arctic-Barents Region.

3.2. Human activities and globalisation

As investment in the region grows, human activities also threaten food and human security. In particular, mining, oil and gas, shipping and tourism all challenge food security in the Barents.

The timber industry has a significant impact on traditional food security, and sustainable forestry is vital to ensure the continuous promotion of food availability and accessibility for both local and indigenous communities. The forests are not only used for herding reindeer but all across the Arctic-Barents, they are also a great source of traditional foods such as berries, herbs, shrubs, lichen, moose, reindeer, lynx and wolverine. Both indigenous and non-indigenous peoples use the land and environment for their traditional hunting, gathering and cultural practices.

Reindeer husbandry has a special position among indigenous peoples, especially amongst the Sámi and Nenets in the Barents Region (WGIP 2017). Herding areas consist of coniferous forests that are heavily exploited by the forestry industry (AWRH 2016). Logging threatens such practices, as reindeer are herded to specific locations in the forest to graze, especially during the winter months, and depend on the forest for food such as lichen or tree-hanging lichen (Lawrence and Raitio 2006, 1–2). Because of exploitation, sources of food for reindeer and the availability of non-wood forest products are becoming increasingly limited.

Industrial activities and motorised vehicles pose various hazards to nearby animals and species, leading to a loss in wildlife habitat. Noise can be potentially disruptive to wildlife, causing nearby species to migrate further, and forcing hunters to travel a greater distance or find new hunting grounds altogether (Green and Giese 2004, 85). In this way, the availability and accessibility of traditional foods is threatened. In addition, chemicals and heavy metals entering the water supply from heavily industrialised areas pose a risk for animals and humans alike. In the town of Nikel in Russia, for example, pollution has had a drastic effect on nearby local wildlife, the environment and even food sources such as berries and mushrooms (Hansen 2016). The town's pollution has been at the top of the environmental agenda of the Barents Cooperation (Nilsen 2016). Ultimately, the continued utilisation of contaminated food and water, despite the risks associated with their consumption, pose serious threats to health.

The impact of globalisation and the growing roles of multinational corporations associated with extractive industries is also a challenge to food security in the Barents Region. In order to maintain a sustainable Arctic-Barents Region, these multinational corporations need to ensure their extraction procedures do not pollute or contaminate the pristine land and water of the region (Röver 2014, 3). In addition, oil and gas are established industries by multinational corporations in the Barents Region. Norway and Russia are world leaders in the exploration and exportation of both products. The World Wide Fund for Nature (WWF) has expressed concerns regarding oil and gas exploration over ineffective cleanup methods in instances of oil spills during industrial oil and gas activities. These activities pollute the water and the land meant for fishing and farming respectively thus affecting the availability and utilisation of fish and crops. Noise pollution from drilling and production techniques can also be a problem, they can disturb and cause acoustic damage to fish and other animals using sound to navigate the ocean for food, and in turn impact on both the economic and food value of such fish for the community (WWF 2016). As mentioned above, changes to the ecosystem resulting from human activities can affect all four dimensions of food security in the Barents communities and their waterways.

As the Barents Sea gets warmer, coastal areas used for fishing and hunting will be negatively impacted by a surge in Arctic shipping. Increasingly warming sea ice consequently means disruptions to natural mating and migration patterns for existing marine species, and the displacement of existing species by newer ones (Bogstad 2015). When ships travel more regularly through the central Arctic Ocean, they will likely bring new invasive species to the Arctic as well as to northern ports (Palmer 2013). Shipping is the most common pathway for marine invasive species, and it is responsible for 69% of species introductions to marine areas (Palmer 2013). In turn, both humans and animals that rely on the existing species in the marine ecosystem will be impacted. Hence, the availability and accessibility dimensions of food security are negatively affected.

Amplified traffic could also result in an increase in minor discharges from marine vessels, such as tankers, freighters, fishing boats and coastal ferries. Although these discharges are currently not readily monitored, their impression on Arctic ecosystems may be substantial nonetheless (Nuttall 2000, 1). If people consume fish that have been exposed to these minor discharges, they face health-related risks, and therefore amplified traffic is a threat related to food utilisation for coastal communities in the Barents Region.

Tourism is a growing industry in the Barents Region, and similar to increased shipping, tourist ferries and ships carry the risk of vessel accidents, spills and pollution discharge, which could have major consequences for the Arctic-Barents environment. As previously mentioned the noise from motorised vehicles such as helicopters, snowmobiles and planes for recreational purposes disturbs animals and can have a negative impact on the availability and accessibility dimensions of food security.

Overall, a surge in tourism means a rise in garbage and waste, exclusively in an area where decomposition is slow and waste remains visible on top of the permafrost in many areas (Huntington et al. 2001, 102, box 36). This garbage and waste might eventually end up in rivers; when eaten by fish they enter the food chain, which results in unsafe foods. When humans eat this food is can result in poor health.

In the Arctic-Barents Regions, sports fishing and hunting have also gained popularity amongst tourists, but it frequently puts pressure on resources and leads to conflicts between local and visiting hunters (Snyder 2007, 15). There can also be a conflict of interest between biological diversity and cultural diversity, as exemplified in the case of State governments and the indigenous Sámi communities on salmon fishing (Ween and Colombi 2013, 479). For centuries, the Sámi people have both earned a living and passed on cultural traditions through salmon fishing in the Tana River at the Finnish-Norwegian border. They traditionally rely on drift net and weir fishery, which have been severely restricted by new proposals from the State governments of Norway and Finland to protect salmon extinction (Johansen et al. 2008, 30). As a result, this conflict represents a threat to both Sámi cultural identity and food security.

The overall food system stability in the Barents Region is threatened by human activity, and there is a need for collaborative approaches from all stakeholders in the region to address such threats.

3.3 Environmental pollution

Contaminants can have a severe impact on the health of the environment, animals and people in the region. Contaminants and pollutants are entering the Arctic-Barents Region at an ever-alarming rate due to climate change, increased human activity, globalisation and environmental pollution. Pollutants are transported to the Arctic via long-range atmospheric currents, the marine environment, freshwater/terrestrial routes, and have bio-accumulated in the food chains of individuals and communities living there (AMAP 1998 in Kuhnlein and Chan 2000, 597). Despite the fact that traditional foods are noted as an excellent source of nutrients such as vitamin A, iron, polyunsaturated fatty acids, minerals and energy, they also contribute to good social, spiritual and physical health; they are also the primary source of persistent organic pollutants (POPs) and metals (AMAP 2009, 19).

The Arctic Monitoring and Assessment Programme (AMAP) warned that exposure to environmental contaminants through the traditional diet remains one of the greatest risks to human health in the Arctic (AMAP 2015, 21). Traditional food systems in northern Europe have shown high levels of lead and cadmium in plant ash, fish, sea mammals, birds, water and soil levels (Kuhnlein and Chan 2000, 605).

Exposure to pollutants extends to humans through the consumption of traditional foods, and can have negative impacts on food utilisation. The consumption of certain environmental pollutants can adversely affect the development of the immune system (AMAP 2015, 97). The AMAP 2009 Report on Health explained that, 'diet is the single most important predictor of contaminant exposure in the Arctic populations'. The daily decision for people choosing between what is safe and nutritious can be a daunting task. Henceforth, there is a clear concern about contaminants, cultural values and the availability of traditionally hunted species due to climate change, which all play a role in influencing the types of traditional foods consumed, the frequency of their consumption and the exposure of Arctic-Barents populations to contaminants (AMAP 2009, 19).

Traditional food in the Arctic-Barents includes beliefs about food health benefits and spiritual provisioning, use of food for its educational value, economic benefits and place in the social fabric of community life (Kuhnlein and Chan 2000, 617). Many researchers have noted the change in diet away from traditional foods of meat and fish towards a more market-based diet (Haraldson 1962; Kozlov et al. 2008, and Nilsson 2012, 12–13). The reasons for this change has been associated with the lack of or lower quantities of traditional food resources and increased commercially produced market foods (Kuhnlein and Chan 2000, 598). This trend was observed in Sámi communities, and, with the warmer climate becoming less comfortable to them, has affected their livelihoods and health (Nuttall et al. 2005, 679).

According to AMAP, the changes in food sources can be influential in the nutritional quality, density and food security of indigenous and non-indigenous peoples (AMAP 2009, 21). The report further stated that, nutritionally speaking, the problem is not with the imported food itself, but rather the widespread replacement of traditional food by a diet that is high in sugar and other foods with low nutrient density (AMAP 2009, 21). A shift away from traditional foods, towards store-bought or market-based foods, threatens the nutritional value and health benefits from these foods; it also leads to scarcity and threats regarding traditional seeds, knowledge on plant foods that are used as medicines, food animals and the cultural practices associated with their protection and survival (Egeland and Harrison 2013, 20). In fact, consumption of market-based foods has been linked to an increase in obesity, diabetes and cardiovascular disease in many Arctic-Barents communities (McAuley and Knopper 2011, 1). Unlike traditional wildlife foods, however, market food sources are monitored for contaminants through domestic policies and international trade laws; to this extent, there is some assurance that food species and ingredients are safe for consumption (Kuhnlein and Chan 2000, 616). In such cases, imported or market foods can provide some level of increased availability, accessibility and safer utilisation of food as suitable alternatives in some cases. Overall, the effects of globalisation limit the capacity of local communities to take active roles in promoting their traditional foods and ensuring food security.

4 Conclusions

Food security in the Barents Region has been relatively under-researched, making it difficult for policy makers to accurately assess and address the needs of indigenous and non-indigenous peoples in the region.

It is clear that the impacts of climate change, coupled with human activities and globalisation, penetrate every aspect of life for people in the Barents Region. These changes are occurring at a rapid pace, and impact human security and related intertwined securities such as health, social, economic and community. Therefore, a food system that responds to the threats posed by climate change, human industrial activities and globalisation is required in order to ensure food security in the region. Changes to livelihood patterns have affected how foods are produced, processed, distributed and consumed. There is a need to collect and monitor data on food consumption patterns progressively, from both traditional and imported food products, to determine the potential health consequences from contaminants. Duhaime and Bernard (2002, 37) suggested the development of scenarios to provide decision makers with the information and data required to orient their decisions towards the sustainability of economic systems, the sustainable exploitation of resources, the optimisation of social health conditions, and the preservation of community food security. In addition, ensuring food security in the Barents Region will require adopting effective and specific actions at individual, household, regional and national levels. Strategies should place an emphasis on transforming the agriculture and food systems to mitigate climate change.

The complex and interrelated nature of existing threats in the Barents Region requires an innovative approach to determining solutions for food security. Two ways by which potential solutions can be provided in the future include the need for a better cross-border collaboration and knowledge sharing amongst the stakeholders. Secondly, opportunities for food business enterprises in the region that will encourage a 'niche' that add value to these traditional foods.

References

AMAP. 2009. *AMAP Assessment 2009: Human Health in the Arctic. Arctic Monitoring and Assessment Programme (AMAP)*. Oslo, Norway.

AMAP. 2015. *Arctic Monitoring and Assessment Programme. AMAP Assessment 2015: Human Health in the Arctic*. Oslo, Norway.

AWRH. 2016. Sami & Finns – Finland. Reindeer herding: a virtual guide to reindeer and the people who herd them. http://reindeerherding.org/herders/sami-finns-finland/.

BEAC. 2016. Facts. *Barentsinfo.org*. www.barentsinfo.org/Barents-region.

Berg, E. 2014. Sami traditions: Márkomeannus contribution to the revitalization of Sami food traditions. Master's thesis, UiT Norges arktiske universitet, The Arctic University of Norway, November.

Bertozzi, L. 1998. Tipicidad alimentaria y dieta mediterranea. In A. Medina, F. Medina and G. Colesanti (Eds.), *El color de la alimentacion mediterranea. Elementos sensoriales y culturales de la nutricion* (pp. 15–41). Barcelona: Icaria

Bogstad, B., Gjøsæter, H., Haug, T. and Lindstrøm, U. 2015. A review of the battle for food in the Barents Sea: cod vs. marine mammals. *Frontiers in Ecology and Evolution*, 25 March.

Burek, K.A., Gulland, F.M.D. and O'Hara, T.M. 2008. Effects of climate change on Arctic marine mammal health. *Ecological Applications* 18(2): 126–134.

Burke, M. and Lobell, D. 2010. Climate effects on food security: an overview. In *Climate Change and Food Security*, edited by M. Burke and D. Lobell. Netherlands: Springer.

Dudarev, A.A, Dushkina, E.V., Sladkovy, Y.N., Alloyarov, P., Chupakhin, V., Sladkova, Y.*et al.*2013. Food and water security issues in Russia II: water security in general population of Russian Arctic, Siberia and Far East, 2000–2011. *International Journal Circumpolar Health* 72.

Duerden, F. 2004. Translating climate change impacts at the community level. *Arctic* 57(2): 204–212.

Duhaime, G. and Bernard, N. 2002. Regional and circumpolar conditions for food security. In *Sustainable Food Security in the Arctic: State of Knowledge*, edited by G. Duhaime. Edmonton: CCI Press.

Egeland, G.M. and Harrison, G. 2013. Health disparities: promoting Indigenous Peoples' health through traditional food systems and self-determination. In *Indigenous Peoples' Food Systems and Well-being: Interventions and Policies for Healthy Communities*, edited by H.V. Kuhnlein, B. Erasmus, D. Spigelski and B. Burlingame. Rome: FAO.

FAO. 1996. *Rome Declaration on World Food Security and World Food Summit Plan of Action*. World Food Summit 13–17. Rome, November. www.fao.org/docrep/003/w3613e/w3613e00.HTM.

FAO. 2008. *Climate Change and Food Security: A Framework Document*. Rome: Food and Agriculture Organization of the United Nations.

FAO. 2016. *Human Security & Food Security.* Rome: Food and Agriculture Organization of the United Nations.

Forbes, B.C., Kumpula, T., Meschtyb, N., Laptander, R., Macias-Fauria, M., Zetterberg, P., Verdonen, M., Skarin, A., Kim, K.Y., Boisvert, L.N., Stroeve, J.C. and Bartsch, A. 2016. Sea ice, rain-on-snow and tundra reindeer nomadism in Arctic Russia. *Biology Letters* 12: 1–5.

Green, R. and Giese, M. 2004. Negative effects of wildlife tourism on wildlife. In *Wildlife Tourism: Impacts, Management and Planning*, edited by K. Higginbottom. Altona: CRC for Sustainable Tourism and Common Ground Publishing.

Hansen, M. 2016. Heavy metals in food from the Norwegian, Finnish and Russian border region. Norwegian Institute for Air Research, Tromso, Norway. Presentation at Seminar on food security in the Barents Region. Rovaniemi, Finland. August 16.

Haraldson, S. 1962. Levnads- och dödlighetsförhållanden i de nordligaste svenska lappbyarna. *Läkartidningen* 59: 2829–2844. [in Swedish].

Hennessy, T.W. and Bressler, J.M. 2016. Improving health in the Arctic region through safe and affordable access to household running water and sewer services: an Arctic Council initiative. *International Journal of Circumpolar Health* 75(1): 31149.

Huntington, H.P., Kankaanpää, P., Baldursson, S., Sippola, A.L., Kaitala, S. and Zöckler, C. 2001. *Arctic Flora and Fauna: Status and Conservation. Conservation of Arctic Flora and Fauna.* Helsinki: Oy Edita Ab.

Johansen, M., Erkinaro, J., Niemelä, E., Heggberget, T.G., Svenning, M.A. and Brørs, S. 2008. *Atlantic Salmon Monitoring and Research in the Tana River System. Outlining a Monitoring and Research Program for the River Tana within the Framework of the Precautionary Approach.* (Norwegian-Finnish working group on monitoring and research in Tana). Finnish Ministry of Agriculture and Forestry. http://mmm.fi/documents/1410837/1801204/2015_Tenon_lohikantojen_tila_Status-of-the-river-Tana-salmon-populations.pdf/aced31a7-af81-40aa-9c82-9e6147e58134.

Jordana, J. 2000. Traditional foods: challenges facing the European food industry. *Food Research International* 33:147–152.

Kozlov, A., Borinskaya, S., Vershubsky, G., Vasilyev, E., Popov, V., Sokolova, M.*et al.*2008. Genes related to the metabolism of nutrients in the Kola Sami population. *Int J Circumpolar Health* 67: 56–66.

Kuhnlein, H.V. and Chan, H.M. 2000. Environment and contaminants in traditional food systems of northern indigenous peoples. *Annual Review of Nutrition* 20: 595–626.

Lassuy, D.R. and Lewis, P.N. 2013. Invasive species: Human induced. Chapter 16. In *Arctic Biodiversity Assessment, 2013*. Iceland: Arctic Council.

Lawrence, R. and Raitio, K. 2006. Forestry conflicts in Finnish Sapmi: local, national and global links. *Indigenous Affairs* 4. Copenhagen: IWIGA.

McAuley, C. and Knopper, L.D. 2011. Impacts of traditional food consumption advisories: compliance, changes in diet and loss of confidence in traditional foods. *Environmental Health* 10: 55.

Muller-Wille, L., Granberg, L., Helander, M., Heikkila, L., Lansman, A.S., Tuisku, T. and Berrouard, D. 2008. Community viability and well-being in northernmost Europe: social change and cultural encounters, sustainable development and food security in Finland's North. *International Journal of Business and Globalisation* 2: 331–353.

Nilsen, T. 2016. Airborne pollution exceeds federal limits by 10 times. *The Barents Observer*. July 12. http://thebarentsobserver.com/ecology/2016/07/airborne-pollution-exceeds-federal-limits-10-times.

Nilsson, L.M. 2012. Sami lifestyle and health: epidemiological studies from northern Sweden. Umeå University.

Nuttall, M. 2000. The Arctic is changing. Stefansson Arctic Institute and individual authors. Developed in partnership with the EU Raphael Programme. www.thearctic.is/.

Nuttall, M., Berkes, F., Forbes, B., Kofinas, G., Vlassova, T. and Wenzel, G. 2005. Hunting, herding, fishing and gathering: indigenous peoples and renewable resource use in the Arctic. In *Arctic Climate Impact Assessment*. Cambridge: Cambridge University Press.

Paci, C., Dickson, C., Nickels, S., Chan, L. and Furgal, C. 2004. Food security of northern indigenous peoples in a time of uncertainty. In 3rd Northern Research Forum Open Meeting, Yellowknife.

Palmer, L. 2013. Melting Arctic ice will make way for more ships and more species invasions. *Scientific American*, March 6. www.scientificamerican.com/article/melting-arctic-sea-ice-means-more-shipping-and-moreinvasive-species/.

Parkinson, A.J. and Evengård, B. 2009. Climate change, its impact on human health in the Arctic and the public health response to threats of emerging infectious diseases. *Global Health Action* 2(1): 207.

Rees, W.G., Stammler, F.M., Danks, F.S. and Vitebsky, P. 2008. Vulnerability of European reindeer husbandry to global change. *Climatic Change* 87(1): 199–217.

Röver, C. 2014. A Sustainable Arctic: Preconditions, Pitfalls and Potentials. Arctic Sustainability. 3rd Konrad-Adenauer-Foundation Seminar, Bergen, Norway, 8–9 September.

Snyder, J. 2007. Tourism in the polar regions: the sustainability challenge. *UNEP/ Earthprint*.

UNDP. 1994. *United Nations Development Program, Human Development Report 1994*. Oxford and New York: Oxford University Press

UN-Water. 2015. *Water, Food and Energy*. www.unwater.org/water-facts/water-food-and-energy/.

Van Oostdam, J., Donaldson, S., Feeley, M., Tremblay, N., Arnold, D., Ayotte, P., Bondy, G., Chan, L. and co-authors. 2003. *Toxic Substances in the Arctic and Associated Effects: Human Health. Canadian Arctic Contaminants Assessment Report II*. Ottawa: Indian and Northern Affairs Canada.

Ween, G.B. and Colombi, B.J. 2013. Two rivers: The politics of wild salmon, indigenous rights and natural resource management. *Sustainability* 5(2): 478–495.

WGIP. 2017. *Action Plan for Indigenous Peoples in the Barents Euro-Arctic Region (2016–2018)*. Murmansk: Working Group of Indigenous Peoples in the Barents Euro-Arctic Region (WGIP). March. www.barentsinfo.fi/beac/docs/ActionPlan_2017-2018_ENG.pdf.

WWF. 2016. *World Wide Fund for Nature, Arctic Oil and Gas*. wwf.panda.org/what_we_do/where_we_work/arctic/what_we_do/oil_gas/.

2.5 Water security in the Barents Region

Antonia Sohns

Water security permeates all dimensions of security. Equally, biophysical processes, infrastructure, and the dynamic characteristics of political and financial institutions determine water security. This chapter examines water security in the Barents Region and highlights the reciprocal and changing relationship between water security and the diverse dimensions of human security.

1 Definition

The United Nations (UN) defines water security as:

> the capacity of a population to safeguard sustainable access to adequate quantities of acceptable quality water for sustaining livelihoods, human well-being, and socio-economic development, for ensuring protection against water-borne pollution and water-related disasters, and for preserving ecosystems in a climate of peace and political stability.
>
> (UN-Water 2013, 1)

While other definitions of water security vary depending on scale and context, at its essence, water security is understood as the need to balance human water use with water-related ecosystem use (Srinivasan et al. 2017; Wheater 2014; Y. Hossain et al. 2016). The UN definition highlights the four key dimensions of water security that have been detailed in the Arctic context: water availability, water access, utility of water for livelihood and household needs, and future water resource stability (Daley et al. 2014; Penn 2016). Water security is thus recognised as a defining principle of human health and wellbeing (O'Brien and Leichenko 2007; Harper et al. 2011).

The availability of water is determined by assessing the quantity and quality of water that is produced, stored, processed and distributed. Water managers consider a range of factors to determine water availability, from hydrologic conditions and ecological water demands to water use across sectors and water quality indicators. Accessibility, according to the UN, is measured by the percentage of the population with access to 20 litres of drinking water per day that do not contain biological or chemical agents detrimental to health, and are located within a convenient distance from a person's home (UNDP 2000). In rural areas, which describes many Arctic communities, the UN defines a reasonable distance from a home as one that does not require a person to spend a disproportionate amount of time fetching water to meet their needs (UNDP 2000).

Utility is understood as water resources that are sufficient to meet household and livelihood needs and individual preference (Penn 2016; Loring et al. 2016). The utility dimension of water security also captures the risk of water-related disasters, such as extreme weather and flooding in the Barents Region, which may diminish water's usefulness in meeting livelihood needs or damage infrastructure (BEAC 2012). Stability captures the dynamic nature of water, and emphasises that water security is a process, not a state or condition (Penn et al. 2016). Indeed stability is the interaction between utility, access and availability. Future water resource stability is therefore concerned with the sustainable consumption of water supplies over time and water's temporal features, such as seasonal variation, which affect its form and availability.

These four dimensions are also applicable in defining water security in the Barents Region. The concept of water security must be contextualised in the context of indigenous rights to resources, and the traditional water beliefs and practices of the Sami, Nenets and Veps. States in the Barents Region have agreed to cooperate in good faith with indigenous populations, and obtain their consent prior to development, use, or exploration of their water resources (Lávdegoddi 2012). The transformative processes of climate change affect the relationship between water and people in the Arctic. It is becoming ever more important to understand the multiple, interacting factors across different geographic, social, jurisdictional and temporal scales (O'Brien et al. 2007; O'Brien and Leichenko 2007).

2 Contextualisation

The freshwater that flows North atop the North American and Eurasian landmasses and permafrost into the Arctic Ocean accounts for 10 per cent of the total flow from the world's rivers (CliC et al. 2016; Marino et al. 2009). In the Barents Region, these rivers create unique freshwater ecosystems, such as grasslands, wetlands, glaciers, ice sheets and frozen tundra that support diverse species of plants and animals. The more than five million people in the Barents Region depend on these seasonal and changing freshwater resources for survival and for their livelihoods.

Much of the Arctic environment is an arid or semi-arid landscape, with little annual rainfall. An exception to the polar desert environment is the Atlantic sector of the Arctic between Greenland and Scandinavia, which receives more rain from the Gulf Stream (NSIDC 2016). Due to the cold, dry environment freshwater resources are limited in time and place (Alessa et al.

2010). Arctic communities have therefore developed local water sources over thousands of years, such as rivers, ice, snowmelt and lakes and water storage strategies to cope with water's seasonality and changing state (Goldhar et al. 2013). Arctic populations also depend on municipally supplied water delivered by truck or pipe depending on local water availability and context. In the Barents Region, populations rely on piped water supply in cities and also continue to use traditional sources to improve water security due to water system failure, sociocultural perceptions of freshwater, and necessity along herding routes (Alessa et al. 2008b; Alessa et al. 2010; BEAC 2012).

Within each Arctic nation's boundaries there are significant regional differences in water security. Local disparities in access and availability are often obscured by national data (Hennessy and Bressler 2016; Dudarev et al. 2013a). Although there is a lack of data for the Barents Region, to illustrate regional inequalities in water services across the Arctic a 2006 study revealed that 99 per cent of Nunavik, Canada has access to hot running water, compared to only 49 per cent access in Central Chukotka, Russia (Hennessy and Bressler 2016; Poppel and Kruse 2006).

While water access is nearly universal in certain locations, such as the Barents Region, it may be unsafe to drink due to poor water quality, waterborne pathogens or contamination (Hennessy and Bressler 2016; Nilsson et al. 2013b). For example, in the Russian city Nikel, heavy metal emissions and dioxin pollution from the Norilsk nickel plant impaired water quality and led to food safety warnings for certain regional lakes (Inari and Torvinen 2015). These statistics illuminate the insecurities that Arctic populations face in obtaining water of sufficient quantity and quality. This is particularly true in the Barents Region where populations are regularly exposed to pollutants from resource development such as mining (BEAC 2010).

An assessment of water security in the Barents Region and Arctic context should account for its unique local characteristics, such as history and socioeconomic trends. These concomitant factors are dynamic. As the dimensions of water security evolve and change over time, they are coupled with biophysical exposures, such as climate change and seasonality of water resources, which may heighten exposure to biophysical threats (O'Brien and Leichenko 2007). The exposure a community experiences may be compounded by the results of political process and changing economic trends.

In the Barents Region, most of the Nordic area has widespread water purification systems and maintains control of industrial emissions to prevent water contamination, however Troms County in Norway and Västerbotten County in Sweden experience pollution from forestry, agricultural runoff and sewage (BEAC 2010). Compared to the Nordic area, the Russian Barents Region is exposed to severe water insecurity due to unstandardised water protocols, unregulated water quality and poor water supply systems (BEAC 2012; Emelyanova and Rautio 2016). In the Murmansk region, where there is significant mining of copper-ore, nickel and other rare metals, municipal water systems typically only chlorinate drinking water, which may leave harmful, naturally occurring contaminants in the source water (Rautio et al. 2017; BEAC 2012). Indeed, Dudarev et al. found that in up to 51 per cent of drinking water in the northwestern part of the Russian Arctic there were very high levels of chemical contamination (Dudarev et al. 2013a). Cities like Kirovsk, Zapolarny and Nikel, were found to have elevated levels of hazardous pollutants such as aluminium and nickel, and water tested in the Petchenga region had high levels of strontium and barium (Rautio et al. 2017). Most of these water security threats are related to industrial emissions (BEAC 2012).

Insufficient water management, irregular water monitoring and inconsistent and old technology impact the health and wellbeing of the people in the Barents Region (Mueller et al. 2016). For example, the water system in Arkhangelsk Oblast has a leakage rate of 30 per cent during transport of water to customers (BEAC 2010). The aged infrastructure in the Russian Barents Region is representative of the deficient water systems in Russia at-large where more than 27 per cent of water pipes from reservoirs lack water purification facilities and 16 per cent are not equipped with disinfection systems (Dudarev et al. 2013b).

In northern Russian territories, the shortage of good quality drinking water, poor sanitation, unhygienic conditions, decentralised water systems and waste handling have plagued communities (Dudarev et al. 2013a). Such conditions have led to a higher incidence of bacterial dysentery in Yakutia, Taimyr and Chukotka compared to the national average (Dudarev et al. 2013b). Despite identified water concerns in Russia, there is no standard protocol to assess water security and researchers have called for a full reform of the Russian water industry (Dudarev et al. 2013a).

Across the Barents Region, further assessment of the quality and quantity of groundwater is vital to maintaining a sustainable use of water resources and the health of populations, as 95 per cent of North Ostrobothnia County relies on groundwater (BEAC 2010). In several areas of the Barents Region, groundwater quality studies to measure concentrations of heavy metals and pesticides have not been conducted (BEAC 2010). Water security therefore varies depending on the community, and its unique physical, cultural, political and economic characteristics (Alessa et al. 2010; Goldhar et al. 2013). Without improved water management and water monitoring, Arctic ecosystems and communities will be challenged by the changing climate and hydrologic cycle.

3 Assessment

Climate change, human activity and globalisation exacerbate water security across its four dimensions. These challenges are complex and interconnected. The following sections examine some of the impacts these threats have on water resources and future water security in the Barents Region.

3.1 Climate change

The impacts of climate change are manifesting in the Arctic at a startling pace and profoundly affecting freshwater resources and ecosystems. From the small intermittent streams to the intricate networks of shallow ponds and marshes, the Arctic's land and waterscapes are mutable and morphing in the warming north. The changing climate is affecting the stability of water resources and water availability for ecosystems and human populations (McClelland et al. 2004; Prowse et al. 2015; Bintanja and Andry 2017).

Climate impacts on the hydrological regime will affect the volume and seasonality of precipitation and river discharge (Dankers and Middelkoop 2008). With the onset of climate change in the Barents Region, river discharge will increase in all seasons except late spring and early summer (Dankers and Middelkoop 2008; Mueller et al. 2016). These changes alter the availability, utility and long-term stability of water supplies by affecting turbidity, introducing new contaminants and mobilising existing pollutants in the environment. Water availability and utility may be additionally limited due to warmer temperatures and the spread of new species and waterborne pathogens (Martin et al. 2009; Ford et al. 2014; Harper et al. 2011; Harper et al. 2015). Already, the incidence of waterborne diseases in the general population in the Russian Arctic, Siberia and the Far East is very high (Dudarev et al. 2013a).

Communities have observed a gradual drying of lakes and marshes due to higher temperatures, therefore reducing available water supply (White et al. 2007a; Ford and Smit 2004; Martin et al. 2009). The warming temperatures in the Volga and Komi are thawing the land and increasing available water by five per cent, whereas in northern Fennoscandia water in the soil is disappearing (Roderfeld et al. 2008). The majority of changes to soil moisture will take place by 2050, therefore having uncertain yet imminent impacts on the water available to communities across the Barents Region (Roderfeld et al. 2008).

Climatic shifts have increased variability of water resources across temporal and spatial scales and produced new interactions between groundwater and surface water as permafrost thaws (Loring 2010; Martin et al. 2009; Prowse et al. 2007; D. White et al. 2007a). For the people in the Barents Region, the snow season will be shortened by two months and snowmelt will occur earlier (Dankers and Middelkoop 2008; Roderfeld et al. 2008). More precipitation is projected to fall as rain instead of snow, and precipitation is predicted to increase across the Barents Region, thus affecting water supply (Dankers and Middelkoop 2008). Despite more rain and river discharge, heightened variability of the hydrological cycle due to warming temperatures may dry out tundra habitat, alter the timing of spring flooding, and threaten water security (Roderfeld et al. 2008; BEAC 2012).

Water access provided by built infrastructure, including water pipes, wastewater lagoons, and purification systems is being rendered inoperable as they warp in the Earth's active layer. Protective berms that contain wastewater lagoons will be weakened with the thawing ground. This change will compromise wastewater treatment methods in the Arctic, and potentially contaminate drinking water resources with raw sewage (Martin et al. 2009).

Climate change has an impact on water utility, which includes water hazards and vulnerabilities. Storms have damaged and disrupted water and sanitation systems through coastal erosion and storm surge (Brubaker et al. 2011; Thomas et al. 2013). Freshwater supplies in lowland and coastal areas are susceptible to saltwater intrusion from worsening storm surge and rising seas. Such hydrologic changes shift vital chemistry and characteristics of the environment, potentially jeopardising ecosystem services. In the Komi Republic, flooding and deluges have waterlogged the land and threatened homes, power lines and bridges (BEAC 2010). Flooding and the destructive nature of water are of increasing concern across the Barents Region.

The changing stability of Arctic freshwater systems greatly influences human wellbeing. Variable water supplies have serious health consequences for Arctic populations. In the Russian Arctic, poor quality water supplies and a lack of centralised water systems results in an increased incidence of hepatitis A, giardiasis, shigellosis, among other infectious and parasitic diseases (Dudarev et al. 2013a). Despite the proven benefits of sufficient water supplies on human health, health interventions often neglect specific water quality and domestic water supply standards (Daley et al. 2014). Water variability also affects physical health (Goldhar et al. 2013; Daley et al. 2014). Due to the hardship of collecting water, snow and ice, men experience chronic back and shoulder pain (Goldhar et al. 2013; Sarkar et al. 2015) Limited water supply and the fear of water insecurity additionally increase mental stress in individuals (Sarkar et al. 2015).

3.2 Human activities and globalisation

As the Arctic warms, human activity is changing as well. Dynamic shifts such as demographic trends, globalisation, mining, resource development and shipping are increasing water insecurity in the Barents Region. In particular, the mining industry is growing and its impacts on water resources are widespread. Mining operations contaminate local water resources with tailings and waste, thereby reducing local water availability and utility (BEAC 2012; Sandlos and Keeling 2016). Leaks during mining operations, such as in the Terrafame mine in Sotkamo, eastern Finland and in Russia by Norilsk Nickel have resulted in significant contamination of surface and groundwater with elevated levels of metals including nickel, uranium, aluminium and copper (News 2017; AFP 2016).

Abandoned mines also leach toxic waste into the environment and contaminate water resources, adversely impacting human health (Sandlos and Keeling 2016). In Västerbotten County some industrial sites risk contaminating water bodies with heavy metals (BEAC 2010). With the onset of climate change, more extreme weather will challenge existing mining operations and may result in more leaks. These water releases pollute nearby water resources and threaten the health of those who rely on local water supplies (Pearce et al. 2011).

Additionally, a less predictable water supply may impair the energy industry and its operations. Changing seasonal precipitation patterns and higher temperatures will limit water availability for operations and will reduce recovery rates further. In many cases, mining is heavily dependent on local water supplies and increased water scarcity can threaten a mine's rate of production, its dust suppression ability and mine drainage. Despite the risks and potential climate impacts, mining operations and regulations have remained largely unchanged. Some mining operators have expressed concern regarding the lack of adaptation planning for mining operations due to uncertainty around climate change impacts (Pearce et al. 2011; Ford et al. 2010).

In the case of power generation, extreme weather will impact the planning and operation of energy systems (Snorrason and Jonsdottir 2005). Hydropower will be especially affected by climate change as precipitation regimes and temperatures shift, thereby altering the annual inflow of water and the seasonal distribution of water (Golombek et al. 2012). Such variations impact the reliability of hydropower to supply electricity due to fluctuating water levels in the reservoirs (Golombek et al. 2012; Rodriguez et al. 2013). It is predicted that increased precipitation and river runoff will benefit electricity production from hydropower (Lehner et al. 2005; O'Brien et al. 2004). However, as river discharge is anticipated to increase by more than 25 per cent in some areas of the Barents Region by 2020, the runoff may increase the magnitude of flooding and therefore damage dam infrastructure (Lehner et al. 2005; O'Brien et al. 2004). Such changes to the hydrologic cycle will not only impact hydropower, but may threaten water quantity and quality for mining, industrial use and power generation (Instanes et al. 2016).

Continued growth of the forestry, oil and gas industries in the Barents Region will additionally challenge freshwater supplies as industry places increasing pressure on water resources for operations. There is global interest in winning mining contracts in the Barents Region. For example, China's state oil company, China National Petroleum Corp, is negotiating a joint exploration of oil resources in the Barents and Pechora Seas with Rosneft, owned by the Russian Government (Larsen and Fondahl 2015). Additionally, changing the land for development, such as through clear-cutting and ditching in forestry in Västerbotten County, mobilises phosphorous and severely diminishes freshwater quality (BEAC 2010). Gas extraction similarly appropriates new land for exploration and development and degrades freshwater resources. On the Yamal Peninsula, development and land use change has impaired water resources to such an extent that water sources may not be able to support tundra nomadism in the future, and herders risk being socially and culturally alienated from the land (Forbes et al. 2009).

Demographic shifts are an additional stress on water security in the Barents Region. Across the Arctic there has been a continued movement towards urban centres, while the smaller settlements are declining in number and size

(Larsen and Fondahl 2015). The growing populations in the Nordic side of the Barents Region increases pressure on local and regional freshwater resources and systems (Medeiros et al. 2016; Instanes et al. 2016; K. Hossain 2016). On the Russian side of the Barents Region, populations have been declining since the disintegration of the Soviet Union (AMAP 2017). Shifting demographics and seasonal fluxes of workers puts additional stress on existing water infrastructure. Additionally, new Arctic shipping routes will bring an influx of tourists. The ships and people may transport new species and damage water resources and freshwater ecosystems in the Barents Region (Larsen and Fondahl 2015; BEAC 2010). Other indirect effects that impair freshwater resources and environments are trash disposal as well as road and infrastructure development (Forbes et al. 2009).

As pollution compromises water resources, cascading impacts across sectors emerge. Forbes et al. (2009) document how degraded water sources due to development result in a loss of fish, and degraded vegetation and soils. Such changes heighten food insecurity for reindeer herders on the Yamal Peninsula (Forbes et al. 2009). These stresses are comparable to other regions around the Arctic, such as Alaska, Canada and Greenland, which are experiencing increasing pressures from resource development and globalisation.

4 Conclusions

Arctic freshwater systems are challenged by aging and insufficient infrastructure, impaired freshwater resources, ecosystem degradation, industrial development, climate change and demographic shifts that are occurring concomitantly (Forbes et al. 2009). Collectively, these factors determine the water security and health of Arctic communities (Nilsson et al. 2013a). To mitigate the impact of these changes on freshwater resources and the water security of the Barents Region, it is vital to increase research on all aspects of water security.

In the Barents Region, there is a great need for more data regarding hydrological characteristics of Arctic rivers and watersheds, and for improved measurement of precipitation (Dankers and Middelkoop 2008). Groundwater assessments and studies should also be conducted to determine the quantity and quality of groundwater resources and to establish a baseline. Measureable quantitative indicators of water security should be implemented and monitored in the Arctic regions over time (Nilsson et al. 2013b).

More robust water governance will improve water security of populations and protect water resources (Dudarev et al. 2013b; Hennessy and Bressler 2016). Due to the international context of the Barents Region, coordinating research and databases will improve regional water security. Consistent language and indicators should be established to measure water security across localities and contexts. The absence of integrated policymaking across sectors in Arctic regions, and by extension the Barents Region, demands urgent attention. Tradeoffs and opportunities between resources should be recognised and reflected in policy. Due to the changing state of the Arctic and the reciprocal vulnerabilities that communities and existing industries will experience, it is vital that Arctic water governance frameworks consider cross-sectoral stressors that will have a lasting impact on water resources today and into the future.

For long-lasting and inclusive water management in the Barents Region, diverse perspectives from around the region must be incorporated into the framework. It is important that strategies to adapt to climate change at all levels of governance are based on understanding the determinants of water security and cultural factors (Nilsson et al. 2013b; Alessa et al. 2008b). There is also a pressing need to identify the implications of freshwater trends on Arctic peoples and their ability to maintain control over water resources with the onset of these imminent changes (Goldhar et al. 2014). More research is needed regarding the impacts of climate change on local freshwater resources, to understand how communities are adapting water and sanitation systems to the warming and changing environment (Cook and Bakker 2012; Alessa et al. 2008b; Alessa et al. 2008a).

With continued research and implementation of strong local and regional freshwater policies, the Barents Region will be able to enhance its water security. In doing so, the Barents Region will be able to improve the health and livelihoods of communities, protect valuable ecosystems and ensure the long-term viability of resource development.

References

AFP. 2016. Small Fine for Russia Metals Giant over Red River Leak. *Phys.org*, October 31. https://phys.org/news/2016-10-small-fine-russia-metals-giant.html.

Alessa, L.N., Kliskey, A., Busey, R., Hinzman, L. and White, D. 2008a. Freshwater Vulnerabilities and Resilience on the Seward Peninsula: Integrating Multiple Dimensions of Landscape Change. *Global Environmental Change* 18(2): 256–270.

Alessa, L.N., Kliskey, A. and Williams, P. 2010. Forgetting Freshwater: Technology, Values, and Distancing in Remote Arctic Communities. *Society & Natural Resources* 23(3): 254–268.

Alessa, L.N., Kliskey, A., Williams, P. and Barton, M. 2008b. Perception of Change in Freshwater in Remote Resource-Dependent Arctic Communities. *Global Environmental Change* 18(1): 153–164.

Alessa, L.N., Kliskey, A., Lammers, R., Arp, C., White, D., Hinzman, L. and Busey, R. 2008c. The Arctic Water Resource Vulnerability Index: An Integrated Assessment Tool for Community Resilience and Vulnerability with Respect to Freshwater. *Environmental Management* 42(3): 523–541.

AMAP. 2017. *Barents Area: Overview Report*. Oslo.

BEAC. 2010. *Environmental Strategy for the Barents Region, WP4. The Interreg III B Baltic Sea*. www.barentsinfo.fi/beac/docs/B2010_environment.pdf.

BEAC. 2012. *Barents Euro-Arctic Council (BEAC) Working Group on Environment. Work Programme for Subgroup on Water Issues*. Finland.

Bintanja, R. and Andry, O. 2017. Towards a Rain-Dominated Arctic. *Nature Climate Change* 7: 263–267.

Brubaker, M., Berner, J., Chavan, R. and Warren, J. 2011. Climate Change and Health Effects in Northwest Alaska. *Global Health Action* 4.

CliC, AMAP and IASC. 2016. *The Arctic Freshwater System in a Changing Climate.* www.amap.no/documents/doc/The-Arctic-Freshwater-System-in-a-Changing-Climate/1375.

Cook, C. and Bakker, K. 2012. Water Security: Debating an Emerging Paradigm. *Global Environmental Change* 22(1): 94–102.

Daley, K., Castleden, H., Jamieson, R., Furgal, C. and Ell, L. 2014. Municipal Water Quantities and Health in Nunavut Households: An Exploratory Case Study in Coral Harbour, Nunavut, Canada. *International Journal of Circumpolar Health* 73.

Dankers, R. and Middelkoop, H. 2008. River Discharge and Freshwater Runoff to the Barents Sea under Present and Future Climate Conditions. *Climatic Change* 87(1–2): 131–153.

Dudarev, A.A., Dorofeyev, V.M., Dushkina, E.V., Alloyarov, P.R., Chupakhin, V.S., Sladkova, Y.N., Kolesnikova, T.A., Fridman, K.B., Nilsson, L.M. and Evengard, B. 2013a. Food and Water Security Issues in Russia III: Food- and Waterborne Diseases in the Russian Arctic, Siberia and the Far East, 2000–2011. *International Journal of Circumpolar Health* 72(1): 21856.

Dudarev, A.A., Dushkina, E.V., Sladkova, Y.N., Alloyarov, P.R., Chupakhin, V.S., Dorofeyev, V.M., Kolesnikova, T.A., Fridman, K.A., Evengard, B. and Nilsson, L. M. 2013b. Food and Water Security Issues in Russia II: Water Security in General Population of Russian Arctic, Siberia and Far East, 2000–2011. *International Journal of Circumpolar Health* 72(1): 22646.

Emelyanova, A. and Rautio, A. 2016. Population Diversification in Demographics, Health, and Living Environments: The Barents Region in Review. *Nordia Geographical Publications* 45(2): 3–18.

Forbes, B., Stammler, F., Kumpula, T., Meschtyb, N., Pajunen, A. and Kaarlejarvi, E. 2009. High Resilience in the Yamal-Nenets Social–ecological System, West Siberian Arctic, Russia. *PNAS* 106(52): 22041–22048.

Ford, J.D. and Smit, B. 2004. A Framework for Assessing the Vulnerability of Communities in the Canadian Arctic to Risks Associated with Climate Change. *Arctic* 57(4): 389–400.

Ford, J.D., Pearce, T., Prno, J., Duerden, F., Berrang, L., Ae, F., Beaumier, M. and Smith, T. 2010. Perceptions of Climate Change Risks in Primary Resource Use Industries: A Survey of the Canadian Mining Sector. *Regional Environmental Change* 10: 65–81.

Ford, J.D., Willox, A.C., Chatwood, S., Furgal, C., Harper, S., Mauro, I. and Pearce, T. 2014. Adapting to the Effects of Climate Change on Inuit Health. *American Journal of Public Health*, Suppl 3: e9–17.

Goldhar, C., Bell, T. and Wolf, J. 2013. Rethinking Existing Approaches to Water Security in Remote Communities: An Analysis of Two Drinking Water Systems in. *Water Alternatives* 6(3): 462–486.

Goldhar, C., Bell, T. and Wolf, J. 2014. Vulnerability to Freshwater Changes in the Inuit Settlement Region of Nunatsiavut, Labrador: A Case Study from Rigolet. *Arctic* 67(1): 71–83.

Golombek, R., Kittelsen, S.A.C. and Haddeland, I. 2012. Climate Change: Impacts on Electricity Markets in Western Europe. *Climatic Change* 113(2): 357–370.

Harper, S.L., Edge, V.L., Ford, J., Willox, A.C., Wood, M., McEwen, S.A., Alley, R. 2015. Climate-Sensitive Health Priorities in Nunatsiavut, Canada. *BMC Public Health* 15(1). BioMed Central: 605.

Harper, S.L., Edge, V.L., Schuster-Wallace, C.J., Berke, O. and McEwen, S.A. 2011. Weather, Water Quality and Infectious Gastrointestinal Illness in Two Inuit Communities in Nunatsiavut, Canada: Potential Implications for Climate Change. *EcoHealth* 8(1): 93–108.

Hennessy, T.W. and Bressler, J.M.. 2016. Improving Health in the Arctic Region through Safe and Affordable Access to Household Running Water and Sewer Services: An Arctic Council Initiative. *International Journal of Circumpolar Health* 75: 31149.

Hossain, K. 2016. Securitizing the Arctic Indigenous Peoples: A Community Security Perspective with Special Reference to the Sami of the European High North. *Polar Science* 10(3): 415–424.

Hossain, Y., Loring, P.A. and Marsik, T. 2016. Defining Energy Security in the Rural North – Historical and Contemporary Perspectives from Alaska. *Energy Research & Social Science* 16: 89–97.

Inari, L. and Torvinen, A. 2015. *Joint Water Management of the Finnish- Norwegian River Basin District (2016–2021)*. http://ec.europa.eu/environment/water/pdf/Finnish_Norwegian_international_river_basin_district.pdf.

Instanes, A., Kokorev, V., Janowicz, R., Bruland, O., Sand, K., and Prowse, T. 2016. Changes to Freshwater Systems Affecting Arctic Infrastructure and Natural Resources. *Journal of Geophysical Research: Biogeosciences* 121(3): 567–585.

Larsen, J.N. and Fondahl, G. 2015. *Arctic Human Development Report. Human Development*. Vol. II. Copenhagen: Nordisk Ministerråd.

Lávdegoddi, Barents Eamialbmogiid. 2012. *Resolution from the 2nd Barents Indigenous Peoples' Congress 2012*. Kirkenes. www.sametinget.se/56722.

Lehner, B., Czisch, G., Vassolo, S. 2005. The Impact of Global Change on the Hydropower Potential of Europe: A Model-Based Analysis. *Energy Policy* 33(7): 839–855.

Loring, P.A. 2010. Ways to Help and Ways to Hinder: Climate, Health and Food Security in Alaska. A Thesis Presented to the Faculty of the University of Alaska Fairbanks. University of Alaska Fairbanks.

Loring, P.A., Gerlach, S.C. and Penn, H.J. 2016. 'Community Work' in a Climate of Adaptation: Responding to Change in Rural Alaska. *Human Ecology* 44(1): 119–128.

Marino, E., White, D., Schweitzer, P., Chambers, M. and Wisniewski, J. 2009. Drinking Water in Northwestern Alaska: Using or Not Using Centralized Water Systems in Two Rural Communities. *ARCTIC* 62(1): 75–82.

Martin, D., Bélanger, D., Gosselin, P., Brazeau, J., Furgal, C. and Déry, S. 2009. Drinking Water and Potential Threats to Human Health in Nunavik: Adaptation Strategies under Climate Change Conditions. *ARCTIC* 60(2): 195–202.

McClelland, J.W., Holmes, R.M., Peterson, B.J. and Stieglitz, M. 2004. Increasing River Discharge in the Eurasian Arctic: Consideration of Dams, Permafrost Thaw, and Fires as Potential Agents of Change. *Journal of Geophysical Research Atmospheres* 109(18).

Medeiros, A.S., Wood, P., Wesche, S.D., Bakaic, M. and Peters, J.F. 2016. Water Security for Northern Peoples: Review of Threats to Arctic Freshwater Systems in Nunavut, Canada. *Regional Environmental Change* 17(3): 635–647.

Mueller, L., Sheudshen, A.K., Syso, A., Barsukov, P., Smolentseva, E.N., Khodzher, T., Sychev, V.G. 2016. Land and Water Resources of Siberia, Their Functioning and Ecological State. In *Novel Methods for Monitoring and Managing Land and Water*

Resources in Siberia, edited by L. Mueller, A.K. Sheudshen and F. Eulenstein, 3–73. New York: Springer International Publishing.

News, YLE. 2017. Terrafame Mine Admits More Widespread Groundwater Contamination, Fights Stricter Environmental Rules. *Eye on the Arctic*, May 9. https://yle.fi/uutiset/osasto/news/terrafame_admits_more_widespread_groundwater_contamination_fights_stricter_environmental_rules/9602479.

Nilsson, L.M., Berner, J., Dudarev, A.A., Mulvad, G., Odland, J.O., Parkinson, A., Rautio, A., Tikhonov, C. and Evengard, B. 2013a. Indicators of Food and Water Security in an Arctic Health Context–Results from an International Workshop Discussion. *Indicators of Food and Water Security in an Arctic Health Context – Results from an International Workshop* 72(1): 9.

Nilsson, L.M., Destouni, G., Berner, J., Dudarev, A.A., Mulvad, G., Odland, J.Ø., Parkinson, A., Tikhonov, C., Rautio, A. and Evengård, B. 2013b. A Call for Urgent Monitoring of Food and Water Security Based on Relevant Indicators for the Arctic. *AMBIO* 42(7): 816–822.

NSIDC. 2016. What Is the Arctic?National Snow and Ice Data Center. https://nsidc.org/cryosphere/arctic-meteorology/arctic.html.

O'Brien, K., Eriksen, S., Nygaard, L. and Schjolden, A. 2007. Why Different Interpretations of Vulnerability Matter in Climate Change Discourses. *Climate Policy* 7(1): 73–88.

O'Brien, K. and Leichenko, R. 2007. Human Security, Vulnerability, and Sustainable Adaptation. No HDOCPA-2007-09, Human Development Occasional Papers (1992–2007), Human Development Report Office (HDRO), United Nations Development Programme (UNDP). https://EconPapers.repec.org/RePEc:hdr:hdocpa:hdocpa-2007-09

O'Brien, K., Sygna, L. and Haugen, J.E. 2004. Vulnerable or Resilient? A Multi-Scale Assessment of Climate Impacts and Vulnerability in Norway. *Climatic Change* 64: 193–225. https://link.springer.com/content/pdf/10.1023%2FB%3ACLIM.0000024668.70143.80.pdf.

Pearce, T.D, Ford, J.D., Prno, J., Duerden, F., Pittman, J., Beaumier, M. and Berrang-Ford, L. 2011. Climate Change and Mining in Canada. *Mitigation and Adaptation Strategies for Global Change* 16: 347–368.

Penn, H.J.F. 2016. Water Security in the Rural North: Responding to Change, Engineering Perspectives, and Community Focused Solutions. Fairbanks, AK: University of Alaska Fairbanks.

Penn, H.J.F., Gerlach, S.C. and Loring, P.A. 2016. Seasons of Stress: Understanding the Dynamic Nature of People's Ability to Respond to Change and Surprise. *Weather, Climate, and Society* 8(4): 435–446.

Poppel, B., and Kruse, J. 2006. Survey of Living Conditions in the Arctic. www.iser.uaa.alaska.edu/Projects/living_conditions/index.htm.

Prowse, T., Bring, A., Mård, J., Carmack, E., Holland, M., Instanes, A., Vihma, T. and F. J. Wrona. 2015. Arctic Freshwater Synthesis: Summary of Key Emerging Issues. *Journal of Geophysical Research: Biogeosciences* 120(10): 1887–1893.

Prowse, T.D., Bonsal, B.R., Duguay, C.R. and Lacroix, M.P. 2007. River-Ice Break-Up/freeze-up: A Review of Climatic Drivers, Historical Trends and Future Predictions. *Annals of Glaciology* 46: 443–451.

Rautio, A., Piippo, S., Pongràcz, E., Golubeva, E., Soloviev, A., Grini, I.S., Altintzoglou, T. and Helgesen, H. 2017. Healthy Living, Nutrition and Food Waste in the Barents Region. Tromsø. https://brage.bibsys.no/xmlui/bitstream/handle/11250/2435199/Rapport%2B05-2017.pdf?sequence=1&isAllowed=y.

Roderfeld, H., Blyth, E., Dankers, R., Huse, G., Slagstad, D., Ellingsen, I., Wolf, A. and Lange, M.A. 2008. Potential Impact of Climate Change on Ecosystems of the Barents Sea Region. *Climatic Change* 87(1–2): 283–303.

Rodriguez, D.J., Delgado, A., DeLaquil, P. and Sohns, A. 2013. *Thirsty Energy*. Washington, DC: World Bank Group. http://documents.worldbank.org/curated/en/835051468168842442/Thirsty-energy.

Sandlos, J. and Keeling, A. 2016. Toxic Legacies, Slow Violence, and Environmental Injustice at Giant Mine, Northwest Territories. *Northern Review* 42: 7–21.

Sarkar, A., Hanrahan, M. and Hudson, A. 2015. Water Insecurity in Canadian Indigenous Communities: Some Inconvenient Truths. *Rural and Remote Health* 15(4): 3354. www.rrh.org.au.

Snorrason, Á. and Jonsdottir, J.F. 2005. Climate, Water and Renewable Energy in the Nordic Countries. In *Regional Hydrological Impacts of Climatic Change – Hydroclimatic Variability, Seventh IAHS Scientific Assembly*. Edited by S. Franks, T. Wagener, E. Bøgh, H.V. Gupta, L. Bastidas, C. Nobre and C. de Oliveira Galvão. Foz do Iguaçu, Brazil: IAHS Publisher. http://iahs.info/uploads/dms/13199.17102–107 Foz S6–2-8 Snorrason.pdf.

Srinivasan, V., Konar, M. and Sivapalan, M. 2017. A Dynamic Framework for Water Security. *Water Security* 1: 12–20.

Thomas, T.K., Bell, J., Bruden, D., Hawley, M. and Brubaker, M. 2013. Washeteria Closures, Infectious Disease and Community Health in Rural Alaska: A Review of Clinical Data in Kivalina, Alaska. *International Journal of Circumpolar Health* 72.

UN-Water. 2013. *Water Security and the Global Water Agenda: A UN-Water Analytical Brief*. Ontario. www.unwater.org/app/uploads/2017/05/analytical_brief_oct2013_web.pdf.

UNDP. 2000. *Access to Safe Water. Charting the Progress of Populations, United Nations Population Division*. www.un.org/esa/population/publications/charting/12.pdf.

Wheater, H.S. 2014. Water Security – Science and Management Challenges. *Hydrological Sciences and Water Security: Past, Present and Future* 366: 23–30.

White, D., Hinzman, L., Alessa, L., Cassano, J., Chambers, M., Falkner, K. and Francis, J. 2007a. The Arctic Freshwater System: Changes and Impacts. *Journal of Geophysical Research: Biogeosciences* 112(G4): G04S54.

White, D.M., Gerlach, S.C., Loring, P., Tidwell, A.C., Chambers, M.C., Aporta, C., BeltaosS. and Burrell, B.C. 2007b. Food and Water Security in a Changing Arctic Climate. *Environmental Research Letters* 2(4): 45018.

2.6 Energy security in the Barents Region

A focus on societal perspectives

Hanna Lempinen and Dorothée Cambou

This chapter discusses the issue of energy security within the broader framework of societal security in the Barents Region. As a 'master resource' (Strauss, Rupp and Love 2013, 11), energy penetrates and crosscuts all aspects of the broadly understood human security: economic, food, health, environmental, personal, community and political security dimensions and concerns (cf. UNDP 1994). However, societal aspects of energy security remain largely understudied and are often overlooked in energy-related debates and decision-making, including in the context of the Barents Region.

1 Definition: the many faces of energy security

There is a wealth of literature discussing the definition of energy security, although the issue of energy remained absent from the 2004 UNDP Report concerning human security. Today, most debates regarding energy security take place through terms which

> denote unimpeded access or no planned interruptions to sources of energy, not relying on a limited number of energy sources, not being tied to a particular geographic region for energy sources, abundant energy resources, an energy supply which can withstand external shocks, and/or some form of energy self-sufficiency.
>
> (Chester 2010, 887)

Traditionally, much of this discussion has revolved around the ways in which these themes play out on the state level in the context of securing 'the uninterrupted availability of energy sources at an affordable price' (IEA 2017). However, along with expanding understandings of 'security' in the broader field of security studies, questions such as security for whom; security for what values; and security from what threat(s) have also entered the stages of energy-related security theorising (cf. Cherp and Jewell 2014).

Indeed, what needs to be acknowledged is the instrumental role that energy has in securing other functions and goals (Scrase and Ockwell 2010): it is not only fundamental to maintaining all political, economic and societal life

(Aalto and Westphal 2007, 5; also Prontera 2009, 9), but also a prerequisite to development of any kind, and essential for everyday activities, quality of life and human survival (Rüdiger 2008, vii; Hemsath 2010, 8; WCED 1987, 141). Echoing the principles of the human security approach, the General Assembly in its 2030 Agenda for Sustainable Development also emphasised the interplay between sustainable development and energy as it envisages a 'world free of poverty, hunger, disease and want … free of fear and violence…with equitable and universal access to quality education, health care and social protection… and where there is universal access to affordable, reliable and sustainable energy' (UN General Assembly 2015). Thus, energy is a core component of human development and wellbeing and cannot be relegated to the spheres of high politics and market transactions nor to the domains of technology and science (cf. Lempinen 2017; Strauss, Rupp and Love 2013). Instead, the all-encompassing nature of energy (security) concerns only underline the importance of engaging lay voices and non-specialist perspectives in the processes of defining what energy security entails (cf. Ciută 2010, 125).

In an attempt to accommodate the foundational importance that energy has for states, societies and communities, some additional dimensions or aspects of energy security have been suggested. Among the most institutionalised definitions is the model of the 'four A's' (cf. e.g. Kruyt et al. 2009), which adds the dimensions of *accessibility* and *acceptability* to the normally included availability and affordability. Out of these, acceptability refers to the awareness over the environmental and societal externalities inevitably associated with all and any energy solutions: meanwhile, accessibility points towards acknowledging the political, economic and technological barriers that might hinder (cost-effective) access to existing energy resources (Fischhendler and Nathan 2014, 153). Bradshaw's (2012) understanding of contemporary energy security is closely related with the four A's and emphasises the resource geographic, institutional, financial and ecological dimensions that together constitute the energy security concern. Out of these, the three former integrate the components of availability, affordability and accessibility, while the latter explicitly adds considerations of ecological and environmental acceptability to the list.

The emphases on the societal and environmental aspects and impacts associated with the energy concern have not emerged in a societal void. There has been a 'growing criticism of cultural ignorance with which many energy implementation projects are handled' (Bastholm and Henning 2014, 1; see also Strauss 2011) and associated concerns over societal acceptance of energy production have received increasing public attention (Mitchell et al. 2011). In a similar manner, the question of accessibility – more often than not conceptualised in state-centric terms – has also become viewed through a more societally attuned lens. Borrowing from the emerging literature on energy justice, questions of access both to energy resources themselves as well as to the benefits derived from their extraction have gradually become included in the domains of energy security debates (cf. Fuller and MacCauley 2016).

As if this conceptual complexity were not enough, the task of defining energy security is only further complicated by the diversity of energy itself: indeed, what 'energy' actually refers to is seldom explicated or explained (cf. Littlefield 2013, 779; Rupp 2013). In order to engage in any meaningful discussion about the energy security situation in the Barents Region, it is therefore necessary to acknowledge the diversity of energy in terms of different energy sources ranging from renewables to non-renewables as well as the specific concerns associated with energy production, transportation and consumption.

Together, the above remarks highlight the inevitably contextual nature of the 'polysemic' (Chester 2010, 887) notion of energy security. Since 'nothing exists that is not energy, or affected by energy' (Ciută 2010, 125), there is a risk that without proper definition and an explicit context, the concept of energy security becomes 'empty' in a sense that it means the '*security of everything* [...] *everywhere* [...] *against everything*' (ibid., 136; italics in original). In this chapter, the required contextualisation is done within the geographical context of the Barents as well as through a focus on the societal aspects that energy security might entail for the peoples, communities and societies within the region.

2 Energy security concerns in the Barents Region

As a part of the broader Arctic region, the Barents has become to be defined mainly in terms of its estimated massive energy endowments and the role they are expected to have in feeding the 'resource-starved world' (Holm 2015, xv). An estimated fourth of the world's remaining hydrocarbon resources are located in the Arctic region, and most of them are expected to be found from the high seas surrounding the Barents (USGS 2009). Indeed, while marine areas are not included in the official definition of the Barents Region, in the specific context of energy they are included in the regional energyscape (cf. Lempinen 2017, 76).

Much like in the context of the broader Arctic, the word 'energy' has become synonymous with the production of oil and gas exports for global markets, and this is true in the Barents Region, which has a long history as a resource exporter (cf. Tennberg et al. 2012, 15–18). The regional energy landscape is, however, significantly more diverse than oil and gas production and transportation. Parts of the region are wealthy in terms of their renewable energy endowments: energy sources such as wind, hydro, tidal, wave, solar and biomass are utilised in and around the region to a varying degree, as is peat energy (cf. Lempinen 2017, 91; Banul 2012). In effect, there are also major differences between the energy supply structures in the different parts of the Barents Region. In general, fossil fuels constitute a major energy source in all Barents parts but only dominate in the energy mix of the Russian parts. In Sweden and Norway, hydroelectric energy represents the primary energy source (CENTEK AB 2010). In addition, renewable fuels (biomass and urban waste) are important energy sources in Sweden but not in the Norwegian or Russian parts. The significance of nuclear energy and related concerns cannot be downplayed, either: the Murmansk oblast relies heavily on nuclear energy, and Finland has commissioned the construction of a new nuclear power plant in the municipality of Pyhäjoki.

What this production-oriented storyline to a great extent overlooks is that the Barents Region is also a home to around five million local and indigenous inhabitants (BEAC 2016). As in many other parts of the Arctic, long distances and cold climate place some of the residents of the Barents Region amongst the highest per capita energy consumers worldwide (Rasmussen and Roto 2011, 151). The extensive presence of energy-intensive heavy industries only further contributes to the high demand of energy in the region (Glomsrød and Aslaksen 2006, 12). As a result, regional concerns associated with energy efficiency and saving are also acutely important in the overall energy security puzzle of the Barents Region (cf. Hirvaskari and Gerashchenko 2015; also Hemsath 2010, 11, 24).

3 Assessing regional energy security

Against this background, assessing the energy security situation in the Barents Region must equally include the diversity of the Barents as a geographical, political and sociocultural region, as well as the diversity of energy in its own right. Considering that the dimensions and definitions of energy security are still debated, this assessment adopts an approach that combines two of the definitions mentioned above. The definitions of Kruyt et al. (2009) and Bradshaw (2012) deal with energy security in the region through the integrated and overlapping components of availability; affordability; environmental aspects; and acceptability of energy-related decisions and measures in the region. In addition, energy security concerns specifically relating to indigenous people within the region are briefly addressed in a separate, dedicated sub-section.

3.1 Availability

While the concerns of energy availability in the context of energy security have traditionally been conceptualised in state level terms, the importance of reliable everyday energy supplies cannot be underestimated at the level of societies, communities and everyday life. The everyday activities and quality of life in the Barents Region are heavily dependent on – or even 'dictated by' (Hemsath 2010, 8) – the reliable and uninterrupted availability of heat, fuel and electricity. However, this reliability cannot be taken for granted in all parts of the region. In some parts of the Barents, especially in Russia, the unreliable and out-of-date energy infrastructure frequently contributes to delivery disruptions (cf. Lempinen 2017, 92), and the otherwise reliable transmission systems in other parts of the Barents remain vulnerable to unexpected events such as natural disasters (cf. Tennberg and Vola 2014).

As such, both the dependence on long distance transports as well as the overdependence on single energy source pose threats for the availability of energy supplies (Hemsath 2010, 24). However, limitations related to the regional energy infrastructure, as well as the material qualities of different energy sources that the existing energy systems have been designed to utilise, make the diversification of energy supplies in the region a burdensome task (cf. ibid., 9; Lähde 2015; Bridge 2011). While oil and gas still remain the dominant sources of energy, the use of renewable energy is increasingly solicited at the local, national and international levels. In fact, the Barents Region is rich in renewable energy resources, especially its Nordic part, where energy systems have a long tradition of using efficient water, bio-energy, wind and geothermal energy (NORDEN 2010, 11). The Nordic states also have significant potential to increase their production and use of renewable energy, while in Russia, the government has also begun to encourage work on energy efficiency and the development of renewable energy sources since the adoption of the Federal Law in 2009 'On energy saving and energy efficiency improvements'.

However, the required transition from fossil fuel to renewable energy has progressed much further in the Nordic countries than in the Russian parts of the Barents (CENTEK AB 2010). Despite its potential, several barriers still prevents most cities and municipalities in Russia from increasing their energy efficiency and use of renewables. The low price for fossil fuels (CENTEK AB 2010) as well as the lack of awareness, capacities and skills in Russian municipalities in regards to the possibilities and benefits of energy efficiency and renewables constitute major hurdles limiting the development of renewable (Barents Euro-Arctic Energy Working Group 2011). Processes related to the planning and approval of projects also generally limit opportunities for development in the area of renewables. This is notably the case in the area of wind power development projects (NORDEN 2010, 12), which nevertheless continue to flourish in the Nordic countries. Ultimately, increasing the availability of renewable energy in the Barents Region therefore requires new technologies, investments, and policy solutions.

3.2 Affordability

Despite designations of the broader Arctic as the world's new energy province, parts of the Arctic Barents Region are still plagued by energy poverty, despite their tremendous energy wealth. While definitions of energy poverty vary, in its most classical articulations it is defined as a state where more than ten percent of a person's or a household's income is spent on energy and fuel (cf. Boardman 1991). In more qualitative terms, energy poverty has been briefly defined as inadequate access, affordability, reliability and safety of energy resources for consumption (cf. Bazillian et al. 2014, 219–220).

Regardless of the definition adopted, the issue of energy poverty remains a timely concern also for the Barents Region and some of its residents and communities. While in many areas and for many communities energy-related costs do not play a defining role on the everyday level, some findings from the broader Arctic indicate that especially in remote rural communities the high costs of transportation fuels have an impact on subsistence lifestyles and the viability of communities that depend on them (Hemsath 2010). Despite the advantages created by motorised transportation, the dependence of most rural and indigenous inhabitants in the Arctic communities on available and affordable fuel to hunt, fish and gather reindeer undermine their subsistence livelihoods and their capacity to meet their dietary and cultural needs.

However, only considering the theme of affordability in the context of rural and/or indigenous communities is not adequate in covering the whole spectrum of issues related to the economic aspect of regional energy security. The affordability component of energy security intertwines the explicit issue of energy into a broader societal whole. Besides the potential of uncontrollable fluctuations in energy commodity pricing (Hemsath 2010, 24), changes in micro- and macroeconomic situations in the region influence the ability of individuals and households to afford their energy needs (for economic conditions in the Barents Region c.f. Glomsrød and Aslaksen 2008, 11–26). Amongst other things, changes in employment situation, health status or family relations, reforms or cuts in national social security schemes and growing societal inequality can dramatically influence the affordability situation at a household level in otherwise energy affluent communities.

3.3 Environmental aspects

Environmental considerations related to energy production, transportation, and consumption also constitute a large component of the overall energy security situation in the Barents Region from the societal perspective. Many of the 'environmental hot spots' monitored under the Barents cooperation are linked either to the heat and power generation sector, or to handling oil spills and oil or coal mining waste (cf. BEAC 2017, see also chapter 1 on environmental security for further details). In addition to the environmental and climate impacts of energy production, heat and power generation themselves, the risks of spills and leaks associated with the processes of production and transportation pose environment-related energy security threats and challenges. Potential oil spills – both on- and offshore – could have detrimental impacts on the lives and livelihoods within an environmentally sensitive region, such as the Barents (PAME 2009, 136–138; Hemsath 2010, 20–21).

It is important to consider, however, that there is no 'problem-free' means of producing energy that comes without environmental externalities of some kind. Additionally, renewable energy solutions are accompanied by a range of issues to consider, depending equally on the energy source in question as well as the siting of the development project. However, the environmental aspects and impacts of renewable energy alternatives continue to be underestimated and downplayed (for a brief summary see Fischhendler et al. 2015, 198). While recent discussions revolve around wind power developments (Dai et al.

2015, 911), the development of hydropower has also been and continues to be a source of local environmental conflict in some parts of the region (cf. e.g. Strauss 2011, Suopajärvi 2001).

In addition, nuclear energy – a prominent impetus for political cooperation in the Barents Region (cf. BEAC 1993) – continues to pose its own, fundamental concerns in the region. As the existing nuclear reactors in the Kola Peninsula are running on overtime (cf. AMAP 2015, 67) and the construction of a new nuclear power plant has begun in the Finnish Barents (Fennovoima 2017), nuclear energy production in the region will continue for several decades to come. Alongside the existing risks associated with nuclear power plants, potential accidents from floating nuclear power plants or nuclear powered icebreakers, radioactive waste dumped into the Barents and Kara seas during the Soviet times (AMAP 2015, 5–7, 67) as well as issues related to the handling and storage of nuclear waste, continue to pose radiation safety concerns (ibid., 21–33). Furthermore, incidents or accidents in nuclear power plants located outside the geographic borders of the Barents can potentially still have repercussions within the region (see Chapter 1 on environmental security).

One crucial aspect relates to the dominant role of oil and gas production to global climate concerns. While climate change remains to a great extent absent from northern energy-related debates (cf. Norgaard 2011; Lempinen 2017, 170–171; Tynkkynen and Tynkkynen, forthcoming), the hydrocarbon-dependent global energy system is the single biggest source of anthropogenic greenhouse gas emissions (Bradshaw 2010, 275). At the same time, climate change 'is the most pervasive and powerful driver of change' (Arctic Council 2016, viii) in the circumpolar north, including the Barents Region, within which fossil fuel based energy is equally produced and consumed in large quantities. In this context, the climate and environmental impacts of regional household energy consumption exemplify a case where the northern residents are not only the referent object of energy security, but whose everyday choices and practices also place them in the position of global energy (in)security providers.

3.4 Acceptability

The acceptability dimension of energy security is heavily involved in weighing the potential gains of energy-related decisions and measures against the harm that they may cause. While much of the discussion over energy activities and their impacts in the broader Arctic revolve around the special challenges faced by the indigenous populations in the region, the position of local, non-indigenous populations as well as the impacts to communities and societies residing outside the Arctic region also contribute to overall concerns (cf. Lempinen 2017, 123–127).

Reports and assessments dealing with the broader Arctic have identified the pervasive and profound ways in which Arctic energy projects penetrate and impact all aspects of northern societies. The impacts of northern energy activities range from macro and microeconomic effects to influencing demography, health, education and training. In addition, they have the potential to impact governance, cultural integrity, contact with nature, and the subsequent interactive effects arising from their intersections (AMAP 2010). As a consequence, no community or society in the Barents Region is immune to the influence of regional energy-related developments.

What constitutes an impact, however, is not as straightforward as it might initially appear. The notion of impact is perspectival in the sense that experienced impacts might not directly reflect the (f)actual and measurable outcomes observed. Instead, the perceptions and impressions of the affected individuals and communities are always formed in the interplay of their personal histories and 'real-world' developments (Del Río and Burguillo 2008, 1328–1329; Whitton et al. 2015). Furthermore, communities are both different between each other as well as internally heterogeneous. This means that very different views about the energy (security) situation as well as about the ongoing developments might exist both between different communities as well as within one community.

From the perspective of acceptability, the discussion above relates directly to the ethics and morals of the allocation of costs and benefits of energy-related developments in the region (cf. MacCauley et al. 2016) as well as to the issue of local access to the resources produced in the region for the needs of the global markets. Historically, the positive and negative impacts of Arctic energy projects have been unevenly distributed amongst affected communities, between the project operators, and the producing regions (cf. Hemsath 2010, 23; Glomsrød and Aslaksen 2006, 16; AMAP 2010). The acceptability of energy-related developments also converges with broader societal security through the conflict potential that is embedded in any energy-related project relating to elements of experienced injustice and inequality. If the social license to operate (cf. e.g. Gehman, Lefsrud and Fast 2017, Hall 2014) is not secured from the impacted communities, this potential, in turn, can actualise in the form of social unrest both in localised contexts within the region as well as anywhere within networks of global society (Hemsath 2010, 24; Mitchell et al. 2001).

3.5 Energy security and indigenous peoples of the Barents Region

Although the impacts of energy production are broad and play a role in the lives of all inhabitants of the region, indigenous peoples have disproportionately suffered from the effects of energy development and production of energy resources. In addition, the production of coal and uranium mining, oil and gas extraction, nuclear power, hydropower and wind development are bountiful in the Barents Region, and indigenous peoples are among those who benefit least from these developments. As discussed in the chapter concerning community security, indigenous peoples have historically been deprived of access to their traditional land and resources, and despite current progress still suffer from the consequences of colonisation and assimilation policies. Land and territories where indigenous communities live are often resource rich and serve as the location from which companies extract resources.

Today, the development of energy projects from resources on indigenous lands continues to threaten their traditional livelihoods. In the Nordic countries, the Sami way of life, especially in relation to reindeer husbandry, is threatened significantly by competing land uses, often promoted by the governments themselves through natural resource extraction or renewable energy projects (Anaya 2011, 15–21). In Finland for instance, logging activities continue to pose a threat to the Sami livelihoods, in particular to reindeer herding activities. Specifically, this is due to the lack of consideration for the protection of the rights of the Sami people to land and natural resources (Anaya 2011, para. 59). This situation is also comparable to the case of the Sami people living in Norway and Sweden.

Although the recognition of the rights of the Sami people in Norway has improved since the adoption of the Consultation Agreement and the Finnmark Act in 2005, the Sami Parliament of Norway continues to indicate that the Sami people still face a lack of involvement and authority in the decision-making processes affecting their lands and resources. In this regard, the Committee on Economic, Social and Cultural Rights (CESCR 2013) expressed its concerns that Norway's 'measures for the preservation and promotion of Sami culture do not sufficiently guarantee the right of the Sami people to enjoy their traditional means of livelihood (art. 15)' and recommended that Norway should 'take steps to preserve and promote the traditional means of livelihood of the Sami people, such as reindeer-grazing and fishing'. In addition, Sami representatives have also suggested that the implementation of the 2009 Mineral Act 'remains particularly challenging in relation to energy development projects and reindeer husbandry', a statement also supported by the UN General Rapporteur on the rights of indigenous peoples who concluded in 2016 that the 2009 Mineral Act 'does not meet international standards' (Tauli-Corpuz 2016, para. 20–29). In particular, what has been held problematic 'is the lack of specific consultation or consent requirements with respect to the particular Sami communities that will be directly affected by the proposed measures' as well as 'the absence of provisions for benefit-sharing with Sami communities when mines are located on traditional Sami lands' (Tauli-Corpuz 2016, para. 29–31). As a result, the Rapporteur indicated in her conclusion that 'the Minerals Act raises doubts about the State's ability to respect, protect and fulfill human rights in the context of extractive activities' in Norway (para. 32).

The situation of the Sami people in Sweden has also raised many issues. The current Mining Act does not contain any provisions to accommodate any special rights relevant to Sami people, and existing mining policies do not appear to be sufficient in protecting Sami interests and rights over lands affected by mining activities. In her 2016 report, Tauli-Corpuz also concluded that the Swedish regulatory framework does not adequately recognise and protect Sami rights in accordance with international human rights standards (para. 48). In particular, the fact that the environmental code does not require any social impact assessment in order for an exploitation concession to be granted and does not preclude mining activities if negative effects on Sami culture and ways of life are found is considered problematic. Furthermore, the mineral ordinance stipulates that the Sami Parliament has the right to be informed and to express an opinion for exploitation permits, but the regulation does not entail a right to withhold consent to the granting of the permit. This is in contravention to the duty of the state to consult indigenous representatives in order to obtain their consent, as enshrined under the UN Declaration on the Rights of Indigenous Peoples.

There is also much controversy in the Nordic countries regarding proposed wind energy projects and conflicts between Sami communities, governments and industry. Wind energy projects, which are increasing in the northern parts of the Nordic countries to meet renewable energy targets in response to global warming, can cause significant impacts on reindeer husbandry. For instance, the construction of wind turbines can affect the migration patterns of reindeer (Skarin et al. 2015) and consequently undermine the traditional livelihoods of Sami communities. In his 2011 report, the Special Rapporteur on the Rights of Indigenous Peoples Anaya also raised concerns against the impact of sustainable energy projects, specifically a potential windmill construction boom in northern Norway, which severely affect reindeer calving grounds (Anaya 2011,17). As a result, indigenous populations not only face the impacts of climate change (ILO 2016; Galloway McLean et al. 2009), but also must endure the externalities associated with mitigation solutions. This situates their communities as experiencing greater risk, considering their cultural rights and livelihoods are also at stake (see also chapter on community security). This situation has triggered a number of court cases both in Sweden and Norway between wind energy companies and Sami reindeer herders (NRK 2016; Umeå Court 2015, 2016), which question the compatibility of renewable energy projects and the human rights of the Sami people.

In Russia, Nenets communities face negative economic, social and cultural impacts from the increasing appropriation and reduction of traditional lands used for husbandry, sacred sites and fishing grounds, as a result of oil and gas development projects. According to the BEAC Working Group for Indigenous Peoples, national and international corporations establish activities in areas inhabited by indigenous peoples without involving them in development strategies or planning. In addition, regional administrations often prioritise the development of central areas, at the expense of indigenous communities who live in remote parts of the region (Barents Euro-Arctic Energy Working Group 2017, 12). In this regard, the capacity of energy development projects to meet local communities' needs remains in question. Simultaneously, local communities continue to suffer from a lack of affordable energy and endure the negative effects of its production.

Thus, the survival of indigenous communities relies heavily upon the capacity of the government to balance competing interests in the use of land and territories. However, while the protection of the rights of indigenous peoples' has made progress in the Barents in recent years, it does not appear to be sufficient in ensuring the protection of their rights in accordance with international standards. As a result, there is an urgent need to ensure that the production and use of energy not only meets the needs of indigenous communities but also does not compromise their survival and distinctive livelihoods.

4 Conclusion and recommendations

Energy security is a 'highly context dependent' concept (Kruyt et al. 2009, 2166) whose operationalisation requires a detailed and precise definition. As such, defining energy security within the Barents Region depends not only the energy resource in question but also from the interests, aspirations and experiences of the individuals, families and communities that are affected by or dependent on the energy-related events, decisions and developments. The complexity and diversity of this definition is seldom fully accounted for in the processes of planning and decision-making in the context of energy in the Barents Region. In turn, this has direct implications on the energy security situation as it is experienced by communities – both indigenous and non-indigenous – within the culturally and politically diverse region.

In order to better account for the societal dimensions of energy security, the diversity within environmental externalities associated with energy production need to be acknowledged and addressed. In addition, to better account for the social impacts of energy extraction and infrastructure projects, the manner in which social impact assessments of energy extraction and infrastructure projects – both renewable and non-renewable – are implemented should be improved. In the broader context of extractive industries in the Barents Region, it has been observed that social impact assessments are normally conducted as one-time pre-project assessments in general (cf. Suopajärvi 2015, Kokko et al. 2013). This is done rather than to use them as tools for actually 'analyzing, monitoring and managing the social consequences of development' (Vanclay 2003, 6) throughout the different phases of the energy project. Conducted as they are now, social impact assessments reveal little about the actual impacts of projects on the communities and living environments of the region, as they currently map only the *expectations* that the consulted individuals and communities attach to the expected development project (cf. Suopajärvi 2015). Furthermore, efforts should be made to ensure a more balanced representation within consulted communities (cf. Suopajärvi 2013; Strauss 2011, 51).

In the context of energy production, there is also a need to improve renewable sources. In this regard, it is important for all countries to improve and harmonise regulation and policy actions in order to allow the expansion of renewable regionally. More particularly, it is suggested that the four Barents governments should agree on carrying out coordinated actions in the domain. This would include, for instance, exchanges of statistical data, joint research and development projects, exchanges of experiences between actors, and financial support for public institutions responsible for the promotion of energy efficiency and renewable use of energy (CENTEK AB 2010). Ultimately, increasing the use of renewable resources therefore requires increased regional cooperation at the technological, political, financial and legislative levels.

Alongside issues associated with energy production, more attention should be devoted to local consumption and related concerns. The cold climate, long distances and dependence on energy imports pose significant energy security risks for local communities whose livelihoods, cultural practices and survival depend on the reliability and affordability of electricity, heat and fuel. Measures need to be taken to improve the reliability and safety of existing energy infrastructure and to reduce dependencies on energy imports through a diversification of energy sources and, where possible, through the decentralisation of energy production. Improving energy efficiency and energy saving are also key components of the security of regional energy supplies, as are the measures to address the environmental impacts of energy consumption.

Finally, energy security must also take into account indigenous peoples' rights. The Barents governments, together with indigenous representatives, should consolidate indigenous rights to land and resources to avoid conflicts between energy production and indigenous way of life. At the same time, they should ensure that measure to promote renewable energy development, such as wind projects, do not adversely affect Sami livelihoods. Indigenous decision-making authority and participation in the development of energy project should therefore be improved and consolidated.

However, considering the instrumental and pervasive role of energy across all sectors and all aspects of societal life, a consideration of energy security solely in the explicit context of physical energy sources and production would remain fundamentally inadequate. While regional energy production poses certain threats and challenges within the region, it can also be perceived both as an important contributor to the regional economy and societal wellbeing. It can also be perceived as a source of revenue for states, whose transfer payments heavily support some parts of the northern regions (cf. AHDR 2004, 75; World Bank 2015; on the 'megatrend' of Arctic renewable energy cf. Rasmussen and Roto 2011, 149–168). As such, northern residents are not only dependent on energy itself, but in many instances also from the revenues derived by the states from energy production.

Furthermore, the linkages between fossil fuel production and consumption to climate change should not be overlooked. Climate change remains a major threat to northern cultures, communities and societies, and yet some parts of the Barents Region are heavily involved in and economically dependent on the hydrocarbon industry, which directly contributes to the acceleration of climate change. This profound intertwinement between energy and broader society underlines the urgent need to reconsider the relationships between the global energy system and societal security as a broader whole.

References

Aalto, P. and WestphalK. 2007. Introduction. In *EU-Russian Energy Dialogue: Europe's Future Energy Security*, edited by P. Aalto. Aldershot: Ashgate.

AHDR. 2004. *Arctic Human Development Report*. Akureyri: Stefansson *Arctic* Institute.

AMAP. 2010. *AMAP Assessment 2007: Oil and Gas Activities in the Arctic: Effects and Potential Effects*. Volume 1. Oslo: Arctic Monitoring and Assessment Programme.

AMAP. 2015. *AMAP Assessment 2015: Radioactivity in the Arctic*. Oslo: Arctic Monitoring and Assessment Programme. Anaya, J. 2011. *The Situation of the Sami People in the Sápmi Region of Norway, Sweden and Finland*. UN Doc. A/HRC/*18*/*35*/Add.2

*Arctic*Council. 2016. *Arctic Resilience Report*. Stockholm: Stockholm Environment Institute and Stockholm Resilience Centre.

Banul, K. 2012. Mapping Renewable Energy Policies in the Barents Region from a Multi-Level Governance Perspective. In *Politics of Development in the Barents Region*, edited by M. Tennberg. Rovaniemi: Lapland University Press.

Bastholm, C. and HenningA. 2014. The Use of Three Perspectives to Make Energy Implementation Sutides More Culturally Informed. *Energy, Sustainability and Society* 4(3): 1–14.

Bazilian, M., NakhoodaS. and Van de Graaf, T. 2015. Energy Governance and Poverty. *Energy Research & Social Science* 1: 217–225.

BEAC. 1993. *Kirkenes Declaration*. www.barentsinfo.fi/beac/docs/459_doc_KirkenesDeclaration.pdf.

BEAC. 2016. *The Barents Region*.

BEAC. 2017. *Barents Environmental Hot Spots*. www.barentsinfo.fi/beac/hotspots/.

Barents Euro-Arctic Energy Working Group. 2011. *Energy Efficiency and Renewables; Possibilities and Challenges for Russian Municipalities, Arkhangelsk 11 May 2011*.

Boardman, B. 1991. *Fuel Poverty*. London: Belhaven Press.

Bradshaw, M. 2010. Global Energy Dilemmas: A Geographical Perspective. *The Geographical Journal* 176(4): 275–290.

Bradshaw, M. 2012. Russia's Energy Dilemmas: Energy Security, Globalization and Climate Change. In *Russia's Energy Policies: National, Interregional and Global Levels*, edited by P. Aalto, 206–229. Cheltenham: Edward Elgar.

Bridge, G. 2011. Past Peak Oil: Political Economy of Energy Crises. In *Global Political Ecology*, edited by P. Richard, P. Robbins and M. Watts, pp. 307–324. London: Routledge.

CESCR. 2013. *Concluding Observations on the Fifth Periodic Report of Norway*. UN Doc. E/C.12/NOR/CO/5.

CENTEK AB. 2010. *Preconditions for Improved Efficiency and Increased Use of Renewable Energy in the Barents Region*. Interim report. www.barentsinfo.fi/beac/docs/Energy_efficiency_workshop_Arkhangelsk_11_May_2011_Background_document_ENG.pdf.

Cherp, A. and Jewell, J. 2014. The Concept of Energy Security: Beyond the Four As. *Energy Policy* 75: 415–421.

Chester, L. 2010. Conceptualizing Energy Security and Making Explicit Its Polysemic Nature. *Energy Policy* 38(2): 887–895.

Ciutâ, F. 2010. Conceptual Notes on Energy Security: Total or Banal Security? *Security Dialogue* 41(2): 123–144.

Dai, K.*et al.*2015. Environmental Issues Associated with Wind Energy – A Review. *Renewable Energy* 75:911–921.

Del Río, P. and Burguillo, M. 2008. Assessing the Impact of Renewable Energy Deployment on Local Sustainability: Towards a Theoretical Framework. *Renewable and Sustainable Energy Reviews* 12: 1325–1344.

Fennovoima. 2017. *Hanhikivi 1 Project*. www.fennovoima.fi/en/hanhikivi-1-project.

Fischhendler, I. and Nathan, D. 2014. In the Name of Energy Security: The Struggle over the Exportation of Israeli Natural Gas. *Energy Policy* 70: 152–162.

Fischhendler, I., Nathan, D. and Boymel, D. 2015. Marketing Renewable Energy through Geopolitics: Solar Farms in Israel. *Global Environmental Politics* 15(2): 98–120.

Fuller, S. and MacCauleyD. 2016. Framing Energy Justice: Perspectives from Activism and Advocacy. *Energy Research & Social Science* 11: 1–8.

Galloway McLean, K., Ramos-Castillo, A., Gross, T., Johnston, S., Vierros, M. and Noa, R. 2009. *Report of the Indigenous Peoples' Global Summit on Climate Change.* United Nations University.

Gehman, J., LefsrudL.M. and Fast, S. 2017. Social License to Operate: Legitimacy by Another Name? *Canadian Public Administration* 60(2): 293–317.

Glomsrød, S. and AslaksenI. (eds). 2006. *Economy of the North*. Oslo: Statistics Norway.

Glomsrød, S. and AslaksenI. (eds). 2008. *Economy of the North II*. Oslo: Statistics Norway.

Hall, N.L. 2014. The Discourse of 'Social Licence to Operate': Case Study of the Australian Wind Industry. *AIMS Energy* 2(4): 443–460.

Hemsath, J.R. (ed.). 2010. *Arctic Energy Summit: Final Report and Technical Proceedings.* Anchorage: The Institute of the North.

HirvaskariM. and GerashchenkoI. (eds). 2015. *Multidisciplinary Approach to Develop Energy Efficiency in the Barents Region*. Publication series B. Reports 12/2015. Rovaniemi: Lapland University of Applied Sciences.

Holm, A.O. 2015. Prologue. In *Culture, Development and Petroleum: An Ethnography of the High North*, edited by J.O. Sørnes, D.L. Browning and J.T. Henriksen, xiv–xvii. London: Routledge.

IEA. 2017. *Topic: Energy Security.* www.iea.org/topics/energysecurity/.

ILO. 2016. *Technical Note, Indigenous Peoples and Climate Change From Victims to Change Agents through Decent Work*. www.un.org/esa/socdev/unpfii/documents/2016/Docs-updates/Technical-Note_Indigenous-Peoples_ILO.pdf /.

Kokko, K., Oksanen, A., Hast, S., Heikkinen, H.I., Hentilä, H.L., Jokinen, M., Komu, T., Kunnari, M., Lépy, E., Soudunsaari, L., Suikkanen, A. and Suopajärvi, L. 2013. *Hyvä kaivos pohjoisessa: opaskirja ympäristösääntelyyn ja sosiaalista kestävyyttä tukeviin parhaisiin käytäntöihin*. www.doria.fi/bitstream/handle/10024/96388/Hyv%C3%A4%20kaivos%20pohjoisessa.pdf?sequence=3.

Kruyt, B., Vuuren, D.P., de VriesH.J.M. and Groenenberg, H. 2009. Indicators for Energy Security. *Energy Policy* 37(6): 2166–2181.

Lempinen, H. 2017. *The Elusive Social: Remapping the Soci(et)al in the Arctic Energyscape*. Rovaniemi: Lapland University Press.

Littlefield, S. R. 2013. Security, Independence and Sustainability: Imprecise Language and the Manipulation of Energy Policy in the United States. *Energy Policy* 52(1): 779–788.

Lähde, V. 2015. Politics in a World of Scarcity. In Bergnäs, Kaisa, Teppo Eskelinen. In *The Politics of Ecosocialism: Transforming Welfare*, edited by J. Perkiö and R. Warlenius. Oxon: Routledge.

MacCauley, D., Heffron, R., Pavlenko, M., Rehner, R. and Holmes, R. 2016. Energy Justice in the Arctic: Implications for Energy Infrastructural Development in the Arctic. *Energy Research & Social Science* 16: 141–146.

Mitchell, J., Morita, K., Selley, N. and Stern, J. 2001. *The New Economy of Oil: Impacts on Businesses, Geopolitics and Society*. London: Earthscan.

NORDEN. 2010. *Action Programme for Nordic Co-operation on Energy Policy 2010–2013*. Nordic Council of Ministers. www.barentsinfo.fi/beac/docs/Energy_efficiency_workshop_Arkhangelsk_11_May_2011_Nordic_Action_Plan_on_EP_Cooperation.pdf.

Norgaard, K.M. 2011. *Living in Denial: Climate Change, Emotions and Everyday Life*. Cambridge: MIT Press.

NRK Sápmi. 2016. *Vil at TrønderEnergi trekker seg ut av omdiskutert vindmølleprosjekt*. www.nrk.no/sapmi/vil-at-tronderenergi-trekker-seg-ut-av-omdiskutert-vindmolleprosjekt-1.13369671

PAME. 2009. *Arctic Marine Shipping Assessment 2009*. www.pmel.noaa.gov/arctic-zone/detect/documents/AMSA_2009_Report_2nd_print.pdf.

Prontera, A. 2009. Energy Policy: Concepts, Actors, Instruments and Recent Developments. *World Political Science Review* 5(1): 1–30.

Rasmussen, O.R. and Roto, J. 2011. *Megatrends*. Copenhagen: Nordic Council of Ministers.

Rüdiger, M. 2008. Introduction. In *Culture of Energy*, edited by R. Mogens. Newcastle: Cambridge Scholars Publishing.

Rupp, S. 2013. Considering Energy: E = mc2 = (Magic * Culture)2. In *Cultures of Energy: Power, Practices, Technologies* edited by S. Strauss, S. Rupp and T. Love. Left Coast Press: Walnut Creek.

Scrase, J.I. and Ockwell, D.G. 2010. The Role of Discourse and Linguistic Framing in Effects in Sustaining High Carbon Energy Policy: An Accessible Introduction. *Energy Policy* 38(5): 2225–2233.

Skarin, A., Nellemann, C., Rönnegård, L., Sandström, P. and Lundqvist, H. 2015. Wind Farm Construction Impacts Reindeer Migration and Movement Corridors. *Landscape Ecology* 30(8): 1527–1540.

Strauss, H. 2011. *For the Good of Society: Public Participation in the Siting of Nuclear and Hydro Power Projects in Finland*. Acta Universitas Ouluensis: E Scientiae Rerum Socialium 118. Tampere: Juvenes Print.

Strauss, S., Rupp, S. and Love, T. 2013. Powerlines: Cultures of Energy in the Twenty-First Century. In *Cultures of Energy: Power, Practices and Technologies*, edited by S. Strauss, S. Rupp and T. Love. Walnut Creek: Left Coast Press.

Suopajärvi, L. 2001. *Vuotos- ja Ounasjoki-kamppailujen kentät ja merkitykset Lapissa*. Rovaniemi: Lapland University Press.

Suopajärvi, L. 2013. Social Impact Assessment in Mining Projects in Northern Finland: Comparing Practice to Theory. *Environmental Impact Assessment Review* 42: 25–30. Suopajärvi, L. 2015. The Right to Mine: Discourse Analysis of Social Impact Assessments of Mining Projects in Finnish Lapland in the 2000s. *Barents Studies* 1(3): 36–54.

Tauli-Corpuz, V. 2016. *Report on the Human Rights Situation of the Sami People in the Sápmi Region.* UN Doc. A/HRC/33/42/Add.3.

Tennberg, M. and Vola, J. 2014. Myrskyjä ei voi hallita: Haavoittuvuuden poliittinen talous. *Alue ja ympäristö* 43(1): 73–84.

Tennberg, M., Riabova, L. and Espiritu, A.A. 2012. Introduction to Politics of Development in the Barents Region. In *Politics of Development in the Barents Region*, edited by M. Tennberg. Rovaniemi: Lapland University Press.

Tynkkynen, V.P. and Tynkkynen, N. (forthcoming). Climate Denial Revisited: (Re) contextualising Russian Public Discourse on Climate Change during Putin 2.0. *Europe–Asia Studies.*

Umeå Environmental Court. 2015. DOM Mål nr M 1423–14 Mark- och miljödomstolen 2015-12-10 meddelad i Umeå.

Umeå Environmental Court. 2016. DOM Mål nr M 1425–15 Mark- och miljödomstolen.

UNDP. 1994. *Human Development Report*. New York: Oxford University Press.

UN General Assembly. 2015. *Resolution Adopted by the General Assembly on 25 September 2015. Transforming Our World: The 2030 Agenda for Sustainable Development*. UN Doc. A/RES/70/1.

USGS. 2009. *Assessment of Undiscovered Petroleum Resources of the Barents Sea Shelf. United States Geological Survey Fact Sheet*. http://pubs.usgs.gov/fs/2009/3037/pdf/FS09-3037.pdf.

Vanclay, F. 2003. International Principles for Social Impact Assessment. *Impact Assessment and Project Appraisal* 21(1): 5–11.

WCED. 1987. *Our Common Future*. United Nations World Commission on Environment and Development. www.un-documents.net/our-common-future.pdf.

Whitton, J., Parry, I.M., Akiyoshi, M. and Lawless, W. 2015. Conceptualizing a Social Sustainability Framework for Energy Infrastructure Decisions. *Energy Research & Social Science* 8: 127–138.

World Bank. 2015. *World Development Indicators: Contribution of Natural Resources to Gross Domestic Product*. http://wdi.worldbank.org/table/3.15.

2.7 Personal security in the Barents Region[1]

Tahnee Prior and Patrick Ciaschi

This analysis addresses personal security in the context of the Barents Region. Personal security is one indicator of human security, linked to other forms of insecurity: political, economic, health, food, environmental and community.

1 Definition

Security remains incomplete, unless addressed at all levels. Personal security is one dimension of human security, encompassing both an individual freedom from fear and freedom from want (United Nations Development Programme 1994). The *1994 Human Development Report* regards *personal security* – security from physical violence – as vital for individuals across national contexts. As outlined under the *Human Development Report*, threats to personal security include: threats from the state, like physical torture; threats from other states manifesting as byproducts of war; groupist threats, such as ethnically charged tensions; threats from individuals or non-state actors, such as gang and street violence; threats directed against women, such as rape or domestic violence; and threats directed at children like child abuse, which are based on their vulnerability and dependence.

Personal security, as outlined above, *should* be a priority for *all* states. By definition, the term conjures up the very narrative of 'social contract', construction between states and their constituent citizenries. In exchange for obedience and servitude, new citizens of sovereign states are granted security from outside harm or strife. Today, under the aegis of global liberal governance, the notion of a sovereign state's failure to protect its citizens from human rights violations, ethnocide, genocide and extra-judicial killings can have economic and legal repercussions at the global scale (Muggah 2012).

Compared to other securitisation categories, personal security is a cross-categorical concern (Gasper and Gomez 2015). And it cannot be distilled into a single definition or metric. Individuals and groups respond, adapt and resist, making personal security (as an indicator of welfare and stability) difficult to fully measure. In the context of the circumpolar Arctic and the Barents Region, in particular, personal security poses a number of statistical and technical dilemmas which may pose barriers to effective policy-making and monitoring across the dimensions outlined above.

Moreover, like the concept of human security, personal security is dynamic and diverse depending on the threshold of the individual and concerns multiple overlapping processes of securitisation. It captures both real and perceived threats to individuals (the experiential scale) and the trials and tribulations of localised life. It is thus experienced and framed differently by different groups. Threats to personal security emerge in sub-national contexts of gender inequality or extreme poverty, and do not map onto statist legal conceptions so neatly. Due to the subjective quality of personal security, the collection of data pertaining to personal security is scarce and at times outright problematic (Buzan et al. 1998; Buzan and Hansen 2009; Muggah 2012).

To many critics, the concept is overused, full of 'hot air' (Paris 2001), and void of empirical groundedness. Nevertheless, the concept still holds heuristic value due to its polemical and highly politicised use. It also touches on the delicate subject matter of state and non-state actors' 'responsibility to protect', and the thorny, yet evolving, definition of who are the legitimate actors of care and welfare in global politics.

2 Contextualisation

The Arctic is changing at an unprecedented rate (see for example Arctic Council and Stockholm Environment Institute 2016; Rosen 2017). While traditional security issues, like war and conflict, may not be a source of tension in the region, non-traditional security issues pose significant challenges for its inhabitants. Across the circumpolar north, including the Barents Region, personal security is increasingly threatened by climate change, developments in resource extraction, and in- and out-migration.

In this chapter, we focus on three particular dimensions of personal (in) security: suicide, domestic violence, migration and mobility. In exploring some of these areas, we place a focus on two groups in particular: women and indigenous peoples. While other marginalised groups – such as children, disabled persons, the elderly, racial and ethnic minorities, as well as sexual minorities – experience personal insecurity as well, indigenous peoples and women provide a compelling case; making the complexity of various socio-cultural distinctions and power dimensions particularly visible. One becomes acutely aware of a dearth of data, despite political commitment to both groups in the Barents Region.

In its 1994 *Human Development Report* (United Nations Development Programme 1994, 22), the United Nations highlighted that women generally experience the worst personal threats to security, arguing that 'personal insecurity shadows [women] from cradle to grave.' Women face ongoing disadvantages in their access to economic and social resources in both developing and developed countries (United Nations Development Programme 2007; Nilsson et al. 2004). These resources include access to land, financing, new technology, bargaining power, social capital and training in climate adaptation and disaster risk reduction (Prior et al. 2013; United Nations Development Programme 2011). Indigenous peoples, too, bear the brunt of personal insecurity, yet are not taken seriously as securitising actors. Claims and interests by indigenous communities, such as the Nenets or Sámi, particularly in the realm of environmental or social security, are not always articulated vis-à-vis the language of securitisation (Greaves 2016). As such, their claims are missed, or ignored.

Although all circumpolar states, aside from Russia, maintain a high level of development, this does not necessarily guarantee a high level of gender or racial equality, too (Lahey et al. 2014). In the northern regions of the eight Arctic states, including the Barents Region, there is also a clear difference in human development and gender equality. Human development, gender inequality and aboriginal human development indices, produced by the United Nations, rank states based on a composite statistic that includes (among others) life expectancy, education and labour market participation.

Studies isolating economic development as an indicator of a country's level of human security elide gender or racial inequality from their analyses. Formal rights also do not necessarily guarantee that indigenous women will not continue to face systematic violations, deepened exclusionary and discriminatory practices along gendered and ethnic lines, both within their own peoples and in Arctic society generally.

To avoid such pitfalls in our assessment, we apply an intersectional lens to the three dimensions below: suicide, domestic violence, migration and mobility. An intersectional lens accounts for historical, social and political context and recognises the unique experience of the individual (see Aylward 2011). In doing so, it illuminates a combination of multifarious oppressions which combine to produce a unique and distinct form of discrimination (Eaton 1994, 229). In the context of the Barents Region, an intersectional lens highlights substantial diversity in people's lived experiences.

Governance institutions in the Barents Region, like the Barents Euro-Arctic Council (BEAC), recognised the importance and intersection of gender equality and minority rights early on in Arctic cooperation. At the first meeting of the Barents Forum, in Kirkenes in 1997, and the first conference, in Alta in 1999, gender equality was on the agenda (Barents Euro-Arctic Council 2009). The 2007 Joint communiqué of the Barents Euro-Arctic Council, too, highlighted the importance of 'gender equality in the Barents Region and the strengthening of women's, including indigenous women's, employment opportunities, entrepreneurship and public representation' (Barents Euro-Arctic Council 2007). At the 2013 Barents Summit – the 20th Anniversary of the BEAC – Prime Ministers and other high-level representatives signed a declaration which highlighted the importance of gender equality as key to 'economic activity, growth, and prosperity' (Barents Euro-Arctic Council 2013). Finally, in its 2015 Joint communiqué, XV session, the BEAC recognised that 'a balanced integration of the economic, social and environmental dimensions and respect for democratic values, non-discrimination, gender equality and the rights of minorities, incl. indigenous peoples, are prerequisites for dynamic, inclusive and sustainable regional development' (Barents Euro-Arctic Council 2015). An analysis of the implementation of such discourse, however, remains to be undertaken. And by incorporating an intersectional lens to our analysis, we are able to locate dimensions of personal security, and how they differ from scaled up forms of societal or communal security, for instance.

3 Assessment

In this section, we focus on three dimensions – suicide, domestic violence, migration and mobility – which can help us pin-point the evolving tensions and conflicts related to personal insecurity as they arise in the Arctic and the Barents Region, more specifically (Greaves 2016; Hoogensen 2014). They can also aid in the elucidation of new social relations emerging out of a new reality in the region: increased border security and refugee movement.

3.1 Suicide

The gendered dimensions of personal security affect both men and women. The changing role of men in the Arctic – attributed by some to a loss of identity, self-worth, social tension, issues of power and control – is having a clear impact on societal security as a whole. As noted in the Arctic Human Development Report (Nilsson et al. 2004, 195), 'economic, identity, and personal insecurities intertwine to create a climate whereby a person's own coping strategies no longer suffice to combat these insecurities.' High rates of suicide among men, with high rates of attempts by women, and violence against women in the Arctic cannot be divorced from historical trauma resulting from colonisation, cultural assimilation or residential schools which is passed down from generation to generation (Larsen et al. 2015, 333; Nilsson et al. 2004, 195).

The Barents Region is no exception. Health statistic indicators dating back to 1998 indicated that suicide was much more prevalent among Russian than Nordic men (NOMESCO 1998). Women also exhibited significantly higher rates of suicide when compared to those living in the same state, but south of the Arctic Circle. Suicide rates among young Finnish women, in particular, were considerably higher when compared to other parts of the Nordic Barents Region (NOMESCO 1998). However, the dearth of aggregate and up-to-date data on suicide rates in the Barents Region makes it difficult to develop comprehensive baselines for the region, as well as assess and compare personal security within and across this context. This can have significant implications for policy-making.

However, research by Sumarokov et al. (2014) found high rates of suicide in Nenets Autonmous Okrug between 2002–2012, with suicide rates at 49.2 (per 100,000 person-years) in the non-indigenous population and 79.8 (per 100,000 person-years) among the Nenets population. Substantially higher suicide rates among the indigenous Nenets population were attributed to a 'lack of 'a sense of indigenous belonging', lack of cultural identity and problems of resilience, being single or divorced, and having lower education' (Emelyanova and Rautio 2016). In their 2015 study of cultural meaning(s) of suicide among indigenous Sámi of Sweden, Stoor et al. (2015) further provide evidence that:

> [the suicides] are strongly associated with reindeer-herding (…) and with identity, also. (…) if you are faced with a choice between to abandon your own identity – that is your life really – to do something else, maybe there won't be much left of yourself.

Their work relates such suicidality to issues of inequality and identity struggle, as well a means of strengthening identity, whereby the act itself falls within socio-cultural normativity, engaging the individuals at hand 'in a fight for their Sámi identity' (Stoor et al. 2015).

For both indigenous and non-indigenous Nenets populations, males showed higher suicide rates than females (Sumarokov et al. 2014). Suicide rates in the rural areas of the Komi republic are particularly gendered, with male suicides outstripping female suicides by nine to one (Emelyanova and Rautio 2016). Based on an amalgamation of three studies, Young et al. (2015) provide that suicide rates among Sámi were equal (ratio 1.27 to 1.27) among males and females in northern Norway; and higher among males than females in northern Sweden (1.17 to 0.76) and Finland (1.78 to 1.26). To the knowledge of the authors, data on the suicide rates of the Veps remains missing.

3.2 Domestic violence

All Barents states are party to the Convention on the Elimination of All Forms of Discrimination against Women. In 1992, the Committee on the Elimination of All Forms of Discrimination against Women adopted general recommendation 19 on violence against women; to 'act to protect women against violence of any kind occurring within the family, at the work place or in any other area of social life' (United Nations and Committee on Economic 2008). Under Article 5(b) of the International Convention on the Elimination of All Forms of Racial Discrimination, Barents states also undertake to guarantee 'The right to security of person and protection by the State against violence or bodily harm, whether inflicted by government officials or by any individual group or institution' (United Nations et al. 1985). Moreover, under Article 1(a) of the Council of Europe Convention on preventing and combating violence against women and domestic violence, Norway, Finland and Sweden have agreed to 'protect women against all forms of violence, and prevent, prosecute and eliminate violence against women and domestic violence' (Council of Europe 2015).

Violence against women – directly and indirectly, through its effect on others – serves as a paradigmatic example of personal security within the broader paradigm of human security (Bunch 2004, 32). Existing research provides evidence of ongoing and potential threats to the roles of indigenous and non-indigenous women in community adaptation and in shaping change in the circumpolar Arctic (Arctic Council and Stockholm Environment Institute 2016). Structural impediments – racial, social and economic – and a spectrum of violence – beating, murder, sexual and emotional abuse – affect circumpolar women (Ministry of Social Affairs and Health 2002) and can also increase women's personal insecurity (Nilsson et al. 2004). A failure to provide adequate resources for women in the circumpolar Arctic is linked to health inequality, low income, high housing costs, homelessness, human trafficking and prostitution (Nakray 2013; Nilsson et al. 2004, 201).

Gender-based violence, in particular, highlights the extent to which structure and agency cannot be divorced from personal security (Nilsson et al. 2004, 195). In the Barents Region, rape and domestic violence are threats to personal security primarily directed against women. The rate of domestic violence experienced by women in Finland and Russia, on a national level, remains high (Nilsson et al. 2004; Piispa 2003; UN Women 2013). For instance, in its concluding observations of the Seventh Periodic Report of the Russian Federation, the Human Right Committee to the International Covenant Civil and Political Rights noted the following with regard to domestic violence (United Nations, Human Rights Committee, and Russia 2015):

> The Committee is concerned about the increase by 20 per cent in the number of reported cases of domestic violence affecting women and children since 2010 and about the slow progress in adopting the draft federal act on the prevention of domestic violence. It also notes with concern the lack of due diligence of law enforcement officers in registering and investigating domestic violence cases, and that support services for victims, including the number of psychological and educational centres and shelters, are insufficient.
>
> (Arts. 2, 3, 7, 24 and 26)

Likewise, Sweden and Norway, have – what can be considered – an uncharacteristically high proportion of 'intimate partner violence' (IPV) (Gracia and Merlo 2016) which runs paradoxically counter to their solidified rankings as two of the top countries under the Gender Gap Index.

Again, research on gender-based violence in the Arctic, and the Barents Region in particular, remains underdeveloped (Nilsson et al. 2004, 195). The development of crisis centres across the Barents Region, in response to domestic violence, represents a partial solution to the issues at hand (Saarinen and Carey-Bélanger 2004; Stuvøy 2014). However, cultural and linguistic understandings of what constitutes a 'crisis' in northwestern Russia (e.g. Murmansk) differ in other areas of the Barents Region (Stuvøy 2014). Crisis centres in the Russian part of the Barents Region, for instance, understand domestic violence as a relational phenomenon, a local human security issue which cannot be divorced from other issues of insecurity (Hoogensen 2014). Since this understanding is at odds with the western funding models that support these centres, the identification and administration of crises, and personal security, is inevitably shifted. Moreover, in the context of indigenous women, violence may not be lesser or larger in extent – although it is (Nordic Council of Ministers 2011) – but its consequences are different in magnitude, potentially including the loss of traditional livelihoods and cultural practices (Østby 2006; UAA Justice Center 2015).

Historically, groups such as the Sámi in Norway have had their cultural definitions of the domestic altered as a result of what Bremmer (2012) calls 'Norwegianisation' ideologies in the post-WWII era. These ideologies reconfigured conceptions and demands on labour, property entitlement and traditional roles of women in Sámi society. Equal partnership found in the social life and livelihoods of Sámi society was in many cases eroded in favour of the Sámi woman as the caretaker of the domestic domain and a passive subject of political, economic and social life. Kuokkanen (2006), likewise, sees current forms of domestic violence and gender inequalities in the household as founded upon Christian and Laestadianism conversions and colonisations over the time period of generations, which have redefined the very essence of Sámi women and created a moral hierarchy predicated upon gendered categories. While changes in the domestic sphere are part and parcel of the *longue durée* violence of colonisation, Kuokkanen's (2014) later work also reminds us that externalising violence – as a reflection of the victimhood or loss of status of indigenous men – denies agency and condones the behaviour perpetrators, thereby failing to 'account for the internalisation of patriarchy, which perpetuates the colonial construction of indigenous women as second-class citizens and subordinate members of their communities.' With such intersectional analysis, it becomes evident that domestic violence evolves and intersects with power structures that are nascent, latent and inculcated over the course of generations and thus cannot simply be attributed to one causal source.

3.3 Migration and mobility

Situations of personal (in)security lead to various forms of escape and migration, forced or voluntary in nature. The structural and personal experience of migration conditions livelihoods and is indicative of larger scale shifts in politics and immigration policy alike. Migration is a global governance concern, given the increased movement and displacement of millions of people globally (Office of the United Nations High Commissioner for Refugees 2016). Migration, in its various forms, comes with its own sets of byproducts: permanent loss of home, formal and informal economies, diasporas, the changing social and cultural landscape of communities and fundamental changes in how a 'people' is constituted. Each of these aforementioned effects of migration (as an ongoing fact of human life) generate numerous uncertainties for personal security which transcend the migratory experience. In this section, we focus on three subcategories of the migration and mobility umbrella with direct implications on personal security in the Barents Region: 1) in- and out-migration; 2) human trafficking; and 3) the ongoing global refugee crisis.

3.3.1 In- and out-migration

Generally, the circumpolar Arctic (Nilsson et al. 2004) and the Barents Region are home to more men than women. Some Barents communities are also witnessing gender disparities in migration and growing sex-ratio imbalances in population (Emelyanova and Rautio 2016; Autti and Hyry-Beihammer 2014). Historically, the ratio of men to women has been disproportionate, with more men living in the region. The same holds true in Arctic indigenous communities, where the number of men outstrips the number of women. This is, in large part, due to the out-migration of young females and the in-migration of men for employment in northern extractive and fishing industries (Larsen et al. 2015, 50). While the Nordic states remain closer to the global gender ratio average, the number of women exceeds the number of men in the Russian Arctic (990 men per 1000 women) (Larsen et al. 2015, 69–70).[2] Such a gender disparity is met with concern across the circumpolar north. Some, like Maria Stenberg, Swedish member of the Nordic Council Citizen's and Consumer Rights Committee, have argued that '[u]nless we get a united political grip on [women's out-migration] there is a great risk that the Arctic will become destitute of women.' The AMAP *Adaptations Actions for a Changing Arctic: Barents Overview Report* (Arctic Monitoring and Assessment Programme 2017, 22), too, points to unbalanced out-migration, based on gender, from the region's rural areas. Further Barents-specific data is unknown to the authors.

Out-migration is often attributed to 'a complex of individual and structural push and pull factors'; fewer opportunities for higher education and employment; structural changes in traditional industries (fishing, herding and farming); structural discrimination; unattractiveness of traditional gender roles for men and women; and gender-based violence (Ingolfsdottir 2011; Nilsson et al. 2004; Sloan 2006; Klishin 2014). Economic development in the region focuses on highly gendered labour markets with incentivisation for male dominated professions: mining, transportation, forestry, and the energy industry (BIPE 2017).

In the face of uneven distribution of capital, women migrants are challenging culturally and gender dominant migration categories – such as the 'economic migrant' – by making the 'personal' (marriage, family) into a cross-border political and social phenomenon of concern (Guild 2013). Personal security, on the one hand, is predicated on robust and stable employment opportunities in circumpolar labour markets, but is simultaneously leading to the evolution of women's migration strategies. A 'welfare gap' – a lack of social security in sending countries as a significant push factor in migration – (Godzimirski and Norsk Utenrikspolitisk Institutt 2005), combined with gender inequality, is leading to sustained survivalist forms of migratory practices on the part of female migrants. The out-migration of 'Russian brides'[3] in the Arctic, is but one example of the individual forms of migration taken by women to circumvent depreciating, yet male dominated, local employment markets (Dudzinska et al. 2015; Godzimirski and Norsk Utenrikspolitisk Institutt 2005).

3.3.2 Human trafficking

Demographic shifts with youth emigration and aging populations (BEAC Ad Hoc et al. 2013) in combination with growing extractive and international shipping industries is linked to the underbellies and dark sides of uneven labour markets and human trafficking (Behr et al. 2013). Human trafficking, in particular, evokes imagery of informal and illicit economies where women are victims of coercive and forced out-migration. While indicative of certain realities, human trafficking – composed of sex trafficking, migrant sex work, child trafficking and general labour exploitation – is a complex industry that coordinates various actors, institutions, legal boundaries and borders which perpetuate and impede its practice, simultaneously.

Like domestic violence, human trafficking is an ongoing international concern and conversation in the Barents Region (Office of the Under Secretary for Civilian Security, Democracy, and Human Rights 2016). The Barents Euro-Arctic Council (BEAC) has, at times, actively pursued the issue of trafficking in the region. The BEAC's Task Force for anti-trafficking launched as a result of the 2003 *Kirkenes Declaration*, which emphasised the need for a concerted effort to stop trafficking in human beings in the Barents Region (Barents Euro-Arctic Council 2003). The task force was terminated in 2006 with the reasoning that the length and geographical scope of human trafficking routes were better addressed by the Council of Baltic Sea States (Swedish Ministry of Justice 2006; Barents Euro-Arctic Council 2003). Outcomes of such conversations include the establishment of the 'Nordic Model', which criminalises and aims to prevent sex trafficking in the Barents Region and the Nordic states more generally. With gender equality and anti-exploitative foundations as the model's lynchpin, the states of the Barents Region represent leaders on the anti-trafficking front and the model has been deemed successful on multiple fronts (Skarhed 2010). Nevertheless, trafficking and prostitution have not been eradicated. Analyses argue that, despite the introduction of anti-prostitution laws in Sweden in 1999, both trafficking and prostitution economies in the country remain relatively stable. In many cases, they have merely been 'invisiblised' (Skilbrei and Holmstrom 2016), which points to the multi-faceted problem at hand: legal parameters and crackdowns divert rather than quell the trafficking of women and children.

3.3.3 Global refugee crisis

The Barents Region is also an interesting case for the study of in-migration, or the politics and dynamics of problematised migration forms (i.e. asylum seeking and irregular migration), because there are neither coherent rules, nor a common policy, for migrants to the region (Yeasmin 2013). From a state and transnational perspective on migration governance, the Barents Region poses a number of multi-faceted challenges. Bilateral agreements are in place to facilitate migration between Barents states (i.e. visa-free travel for Russian and Norwegian nationals). These agreements are relatively cooperative and reflective of embedded interdependency within the region. The refugee crisis presents a set of perceived and real problems that change these established relationships and conventions with regards to cross-border migration in the High North. Refugee 'crises' are most notably framed through discourses of fear, and existential threat towards receiving communities (Huysmans 2008). Subsequently, mechanisms for contemporary migration management are either built to control and deter, or granted legitimacy (through plebiscite and referenda) to detain and deport migrants *en masse*. Contrary to the 'Fortress Europe' belief, Nordic states are known as traditional resettlement states (Iceland, Denmark, Sweden, Finland, Norway). Nevertheless, when we compare and contrast the resettlement history across the Barents Region, the narrative is quite different.

At the policy-level, Russia represents an anomaly in the region's more-inclusionary, at least at the level of policy, refugee resettlement track record. Between 2015–2016, roughly 7,000 Syrian citizens have been settled in Russia, and around 5,000 asylum seekers have crossed the Russia–Norway border compared to over 100,000 resettled in Sweden (Chulkovskaya 2016). Aggregate statistics for the Barents Region remain unknown to the authors at this time. Still, these figures do not reflect the turbulences of migratory routes and journeys, however, and do not account for human rights violations: everyday forms of exclusion, intolerance, racism and xenophobia which occur even once refugees are settled in 'inclusive' societies.

In 2016, the Norwegian government, headed by the Progress Party, decided to build a fence on the Russian border, to ostensibly curb border crossings (Osborne 2016). This action is leading to border tensions between Norway and Russia, which is regarded as a migrant transit country in this context (Osborne 2016). Meanwhile, the Finnish government has instituted a new legal technology of border interdiction: the designated country of origin list (DCO), or 'safe country' list. Asylum seeking deterrence is the aim across all: to block the alternative Russian passage into the EU. In 2016, Iraqi, Afghani and Somali nationals were returned to their country of origin, which were considered 'safe' again (Gutteridge 2016).

Regional migration governance, in the context of the global refugee crisis, has been proven ill equipped to manage the flows of people, especially as the Syrian conflict continues. Even though multiple states have adjacent, or contiguous, borders with one another, there are no formal multilateral agreements. There is also little to no harmonised representation of migrants in the region (see Yeasmin 2013), which mirrors how the global governance of migration, and international law towards migration specifically, has 'substance but lacks architecture' (Aleinikoff 2007). Yet, special economic zones, between subregions (think Pomor Zone), exist or are currently in progress. These zones span cross-border territories with their own forms of local visa-free travel.[4] Under increased border controls in the region, these treaties may soon be in jeopardy, due to changing logics or bureaucratic approaches in dealing with documentation and border registrations in this region. More importantly, given the lack of synchronisation on the parts of border regimes, migration routes through Russia and the other Barents states may prove to be more difficult and dangerous for current and future asylum seekers.

On a societal level, the socio-political frame of refugee influx in these rather isolated regions may elevate a discourse of insecurity and shift conceptions of inclusion, tolerance and acceptance. In turn, these sentiments and politicisations could potentially turn into fodder within the public sphere for re-evaluating specific distribution and welfare regimes that refugees rely upon. While personal security here is not as immediate, the effects of ill-equipped resettlement practices and integration models can lead to the discursive construction of 'bogus refugees' through the perceived marginalisations of tax-paying citizens. In her (theoretical and empirical) work on anti-immigrant rhetoric in Finland, for instance, Yeasmin (2016) points to a lack of studies that examine the interaction of this construction with positive liberal attitudes toward international investment and immigration. It remains to be seen whether or not shifts in right wing policy and public reactions towards further refugee resettlement will have a negative impact on personal conceptions of security on the part of individual citizenries.

4 Conclusion

Attempts by state and non-state actors alike to alleviate, and in some cases eliminate, threats to personal security in the Barents Region have proven difficult to fully assess. Direct causal links are difficult to make, data is often scarce, and intersectional and inter-generational lenses are often left unused. These factors all have profound impacts on individuals across spectrums of categorisation, producing the *conditions* of and for insecurity.

In the context of suicide prevention, for instance, additional studies are required to better understand the complex linkages between personal insecurity and identity. As the work of Stoor et al. (2015) provides, suicidality can create a sense of belonging for some. An often-found obstacle is a lack of knowledge of Sámi culture among health personnel in the Barents Region, a barrier to be broken.

Remedial action, too, does not foster immediate positive effects. Gender inequality is not immediately remedied by injections of economic incentives into labour markets (see section on in- and out-migration). Crises centres, seeking to remedy domestic violence, are also faced with cross-cultural governance challenges (see section on domestic violence). Colonisation and the impact of national state modernisation processes over the centuries are also significant factors in the curtailment of gender equalities and the rise of both suicide and domestic violence in indigenous societies. First steps toward truth and reconciliation in Finland, beginning in 2017, provide an opportunity for conversations on ongoing inter-generational trauma and indigenous-state relations to take place. An intersectional approach to this discussion would point to entrenched patriarchal structures, as well.

As Bromfield (2008) makes clear, all Barents states have a notable problem with human trafficking. Despite progressive policies implemented for the protection of non-citizens in cases of human trafficking, the forced migration of women or their labour exploitation remains (see section on human trafficking). Sweden is often upheld as a model country, where its *Aliens Act* provides temporary residence permits to victims of trafficking as of 2007. Still, the challenge of human trafficking remains. Instead of policies of enforcement and policing, which divert or invisibilise even further trafficking routes and economies, improved data collection could effectively point to gaps in law and policy.

In a time of real and perceived crisis (see section on global refugee crisis), bilateral small border traffic deals within Barents sub-regions can be affected by divergent refugee regimes in place. An influx of migrants from outside the region also raises sociological concerns pertaining to integration, multi-ethnic socialisation and religious pluralism; and political concerns on how non-Arctic nationals may alter community life in the North, from a local to a regional scale. The refugee crisis, both discursively and materially, generates new sites of politics in need of both research and policy innovation. Academics have conjured up policy recommendations, and normative frameworks for more efficient, multilateral and embedded mechanisms of migration governance at large (Betts 2011, 2012; Ghosh 2003; Hollifield 2004). While research on trans-regionalist migration within the Arctic – from one Arctic state or city or district to another – is long-standing; research on refugee movement and irregular migration *to* the Arctic from non-Arctic states and the role of such migrants in the Barents Region, in particular, remains an immanent field of study. Still, with the necessary research and data in hand, the Barents Region could serve as a case where forms of meso-level multi-lateral institutions for refugee integration, might be resolved.

While research initiatives like Nordforsk's 'BIPE: Barents International Political Economy: Governance and Gender in Development' (2010–2012) and Arctic Council Sustainable Development Working Group's 'Gender and Equality in the Arctic – Current Realities and Future Challenges' (2015) exist, our assessment points to areas where current work remains insufficient. For instance, sex-disaggregated data and gender-sensitive policies are necessary to protect women's rights to personal security (CEDAW 2009). Yet research and data on gender-based violence remain underdeveloped (Nilsson et al. 2004, 195). Current indicators – from the 2004 and 2014 Arctic Human Development Reports to the Arctic Social Indicators – cannot provide comprehensive gender analysis. Nevertheless, these reports recognise gaps in the development and wellbeing between genders call for gender-disaggregated data, to improve our ability to read a range of other indicators – from food and water to energy security (see Arctic Council and Stockholm Environment Institute 2016, 175; Larsen et al. 2015, 493). Such data would provide a more nuanced understanding of the region's adaptive and transformative capacity, while providing more effective policies tailored to the intersectionality of the Barents' population. What is more, the inclusion of data on women and indigenous peoples in the Barents Region in UN-level data would allow for the development of more comprehensive baselines (Arctic Council and Stockholm Environment Institute 2016, 175; Lahey et al. 2014).

A focus on personal security, in this regard, sheds light on how multiple social, cultural, political and economic areas overlap and affect individual livelihoods. As a result, the complexity of these bridged dimensions provide opportunities for more effective policy creation, and potential multi-layered solutions yet to be considered.

Notes

1 Portions of this paper will be re-published in a forthcoming publication by Tahnee Prior, titled 'Digital Storytelling: A Bottom-Up Approach to Gender & Human Security in the Barents Region?' as a part of an ongoing project, *Human Security as a promotional tool for societal security in the Arctic*. See: www.husarctic.org/.

2 Russia, more generally, has one of the lowest male gender ratios in the world due to high mortality rates and longer female life expectancy (Larsen et al. 2015, 69–70). This ratio is also visible in the Chukotka, where the gender ratio of 105 males per 100 females is even higher than the national ratio.

3 Finnmark, Norway has a significant population of Russian women from Murmansk. 65% of the Russian national migrants in Norway self-identify as female.

4 Travel permits – part of small border traffic agreements (SBTs) – are granted to the residents of the sub-regions comprising the industrial and special economic zones, but are limited to those residents.

References

Aleinikoff, T.A. 2007. International Legal Norms on Migration: Substance Without Architecture. *International Migration Law: Developing Paradigms and Challenges* 467.

Arctic Council, and Stockholm Environment Institute. 2016. *Arctic Resilience Report*. Stockholm: Arctic Council, Stockholm Environment Institute, Stockholm Resilience Centre. https://oaarchive.arctic-council.org/handle/11374/1838.

Arctic Monitoring and Assessment Programme. 2017. *Adaptations Actions for a Changing Arctic: Barents Overview Report*. Oslo, Norway: Arctic Council. www.amap.no/documents/doc/Adaptation-Actions-for-a-Changing-Arctic-AACA-Barents-Area-Overview-report/1529.

Autti, O. and Hyry-Beihammer, E.K. 2014. School Closures in Rural Finnish Communities. *Journal of Research in Rural Education* 29(1): 1–17.

Aylward, C. 2011. Intersectionality: Crossing the Theoretical and Praxis Divide. *Journal of Critical Race Inquiry* 1(1): 1–48.

Barents Euro-*Arctic*Council. 2003. *Barents Euro-Arctic 10 Year Anniversary Declaration*. Barents Euro-Arctic Council. www.barentsinfo.fi/beac/docs/462_doc_BarentsSummitDeclaration.pdf.

Barents Euro-*Arctic*Council. 2007. *11th Session on the Barents Euro-Arctic Council Joint Communiqué*. Barents Euro-*Arctic* Council. www.barentsinfo.fi/beac/docs/JointCommunique11thBEACSession151107.pdf.

Barents Euro-*Arctic*Council. 2009. *Barents Region: Cooperation and Dialogue towards Sustainable Development*. Barents Euro-Arctic Council. www.barentsinfo.fi/beac/docs/Parliamentary_BEAC_info_Eng.pdf.

Barents Euro-*Arctic*Council. 2013. *Declaration on the 20th Anniversary of the Barents Euro-Arctic Cooperation*. Barents Euro-Arctic Council. www.barentsinfo.fi/beac/docs/Barents_Summit_Declaration_2013.pdf.

Barents Euro-*Arctic*Council. 2015. *XV Session of the Barents Euro-Arctic Council Joint Communiqué*. Ministry for Foreign Affairs of Finland. http://formin.finland.fi/public/default.aspx?contentid=335913&nodeid=49549&contentlan=2&culture=en-US.

BEAC Ad Hoc Working Group on Financial, Mechanism Study and BEAC Ad Hoc. 2013. *Financing of Barents Cooperation: Report of the BEAC Ad Hoc Working Group on Financial Mechanism Study*. Helsinki: Ministry for Foreign Affairs of Finland. www.barentsinfo.fi/beac/docs/LOW_UM_Barents_eJulkaisu_A5.pdf.

Behr, T., Brattberg, E., Kallio, J., Aaltola, M., Salonius-Pasternak, C., Raspotnik, A. and Salonen, M. 2013. *The Maritime Dimension of CSDP: Geostrategic Maritime Challenges and Their Implications for the European Union*. European Parliament, Directorate-General for External Policies of the Union. Luxembourg: Publications Office.

Betts, A. 2011. The Global Governance of Migration and the Role of Trans-Regionalism. In *Multilayered Migration Governance: The Promise of Partnership*, edited by R. Kunz and M. Panizzon. Abingdon: Routledge.

Betts, A. 2012. *Global Migration Governance*. Oxford: Oxford University Press.

BIPE. 2017. *Barents International Political Economy: Governance and Gender in Development*. www.nordforsk.org/en/programmes-and-projects/projects/barents-international-political-economy-governance-and-gender-in-development.

Bremmer, M. 2012. *Changing Role of Sámi Women in Reindeer Herding Communities Northern Norway and the 1970–1980s Women's Resistance and Redefinition Movement*. Saarbrücken: LAP LAMBERT Academic Publishing. http://nbn-resolving.de/urn:nbn:de:101:1-201209023066.

Bromfield, N. 2008. Human Trafficking as a Human Rights Issue And Policy Responses of Arctic Nations. In *The Fifth NRF Open Assembly Anchorage*, 1–16. Anchorage. www.rha.is/static/files/NRF/OpenAssemblies/Anchorage2008/a/5th_nrf_anc_2008_footen_bromfield_human_trafficking.pdf.

Bunch, C. 2004. A Feminist Human Rights Lens. *Peace Review Peace Review* 16(1): 29–34.

Buzan, B. and Hansen, L. 2009. *The Evolution of International Security Studies*. Cambridge: Cambridge University Press.

Buzan, B., Wæver, O. and de Wilde, J. 1998. *Security: A New Framework For Analysis*. London: Lynn Rienner Publishers.

CEDAW. 2009. *Statement of the CEDAW Committee on Gender and Climate Change*. CEDAW. www2.ohchr.org/english/bodies/cedaw/docs/Gender_and_climate_change.pdf.

Chulkovskaya, Y. 2016. Why Syrian Refugees Don't Go to Russia. *Al-Monitor*, December 14. www.al-monitor.com/pulse/originals/2016/12/syria-refugees-russia-aleppo-unhcr.html.

Council of Europe. 2015. *Council of Europe Convention on Preventing and Combating Violence against Women and Domestic Violence*. http://site.ebrary.com/id/11021878.

Dudzinska, K., Godzimirski, J. and Parkes, R. 2015. Border and Migration Management in the East: The Cases of Norway and Poland. The Polish Institute of International Affairs. *Policy Paper* 31(133). www.pism.pl/files/?id_plik=20566.

Eaton, M. 1994. Patently Confused: Complex Inequality and Canada v. Mossop. *Review of Constitutional Studies* 1(2): 203–245.

Emelyanova, A. and Rautio, A. 2016. *Population Diversification in Demographics, Health, and Living Environments: The Barents Region in Review*. Nordia Geographical Publications. http://pure.iiasa.ac.at/14468/1/Population%20diversification%20in%20demographics%2C%20health%2C%20and%20living%20environments.pdf.

Gasper, D. and Gomez, O.A. 2015. Human Security Thinking in Practice: 'personal Security', 'Citizen Security' and Comprehensive Mappings. *Contemporary Politics* 21(1): 100–116.

Ghosh, B. 2003. *Managing Migration: Time for a New International Regime?* Oxford: Oxford University Press.

Godzimirski, J.M. and NorskUtenrikspolitiskInstitutt. 2005. *Tackling Welfare Gaps: The Eastern European Transition and New Patterns of Migration to Norway*. Oslo: NUPI.

Gracia, E. and Merlo, J. 2016. Intimate Partner Violence against Women and the Nordic Paradox. *Social Science & Medicine* 157: 27–30.

Greaves, W. 2016. Arctic (In)security and Indigenous Peoples: Comparing Inuit in Canada and Sámi in Norway. *Security Dialogue Security Dialogue* 47(6): 461–480.

Guild, E. 2013. *Security and Migration in the 21st Century*. Oxford: Wiley. http://public.eblib.com/choice/publicfullrecord.aspx?p=1180942.

Gutteridge, N. 2016. Finland Showdown with Brussels after Finns Rule 77% of Iraqi Migrants Could Be Sent Home. *Express.co.uk*, July 22. www.express.co.uk/pictures/galleries/2822/Migrant-crisis-in-Europe.

Hollifield, J. 2004. The Emerging Migration State. *International Migration Review* 38 (3): 885–912.

Hoogensen, G. 2014. *Environmental and Human Security in the Arctic*. London: Earthscan Routledge.

Huysmans, J. 2008. *The Politics of Insecurity: Fear, Migration and Asylum in the EU*. London; New York: Routledge.

IngolfsdottirA.H. 2011. 'Go North, Young Man' – Gendered Discourses on Climate Change and Security in the Arctic. *Nordia Geographical Publications* 40(4): 89–98.

Klishin, I. 2014. Divided Worlds. *The Calvert Journal*. www.calvertjournal.com/features/show/3647/russian-north-murmansk-norway-border.

Kuokkanen, R. 2006. Sámi Women, Autonomy, and Decolonization in the Age of Globalization. *Rethinking Nordic Colonialism*: 1–24.

Kuokkanen, R. 2014. Gendered Violence and Politics in Indigenous Communities. *International Feminist Journal of Politics International Feminist Journal of Politics* 17(2): 271–288.

Lahey, K., Svensson, E.M. and Gunnarsson, A. 2014. Gender Challenges & Human Capital in the Arctic. *Arctic Yearbook*: 1–17. www.arcticyearbook.com/images/Articles_2014/Lahey_AY_2014_FINAL.pdf.

Larsen, J.N., Fondahl, G. and Nordic Council of Ministers. 2015. *Arctic Human Development Report: Regional Processes and Global Linkages*. Copenhagen: Nordisk Ministerråd. http://norden.diva-portal.org/smash/get/diva2:788965/FULLTEXT02.pdf.

Ministry of Social Affairs and Health. 2002. *Taking Wing Conference Report*. Helsinki.

Muggah, R. 2012. Why Personal Security Should Be Part of the Post-2015 Development Agenda. *IPI Global Observatory*, November 13. https://theglobalobservatory.org/2012/11/why-personal-security-should-be-part-of-the-post-2015-development-agenda/.

Nakray, K. 2013. *Gender-Based Violence and Public Health: International Perspectives on Budgets and Policies*. London and New York: Routledge.

Nilsson, A., Larsen, J.N. and Einarsson, N. 2004. *Arctic Human Development Report*. Akureyri, Iceland: Stefansson Arctic Institute. www.deslibris.ca/ID/252336.

NOMESCO. 1998. *Health Statistic Indicators for the Barents Euro-Arctic Region*. Copenhagen: Nordic Medico-Statistical Committee.

Office of the Under Secretary for Civilian Security, Democracy, and Human Rights. 2016. *Trafficking in Persons Report*. US Department of State. www.state.gov/documents/organization/258876.pdf.

Office of the United Nations High Commissioner for Refugees. 2016. *Forced Displacement in 2015*. https://s3.amazonaws.com/unhcrsharedmedia/2016/2016-06-20-global-trends/2016-06-14-Global-Trends-2015.pdf.

Osborne, S. 2016. Norway to Build Border Fence with Russia to Keep out Refugees. *The Independent*, August 25. www.independent.co.uk/news/world/europe/norway-border-fence-russia-refugees-refugee-crisis-schengen-syria-war-a7208806.html.

Østby, L. 2006. Dignity across Borders. In *Dignity across Borders*. Kirkenes: The Northern Feminist University. www.kun.no/uploads/7/2/2/3/72237499/2006_dignity_across_borders.pdf.

Paris, R. 2001. Human Security: Paradigm Shift or Hot Air? *International Security*, 87–102.

Piispa, M. 2003. Violence against Women as Conveyed by Surveys – the Finnish Case. *Journal of Scandinavian Studies in Criminology and Crime Prevention* 3(2): 173–193. doi:10.1080/14043850310013160.

Prior, T., Duyck, S., Heinämäki, L., Koivurova, T. and Stepien, A. 2013. Addressing Climate Vulnerability: Promoting the Participatory Rights of Indigenous Peoples and Women Through Finnish Foreign Policy. SSRN Scholarly Paper ID 2403743. Rovaniemi, Lapland: University of Lapland. https://papers.ssrn.com/abstract=2403743.

Rosen, J. 2017. Arctic 2.0: What Happens after All the Ice Goes? *Nature* 542(7640): 152–154.

Saarinen, A. and Carey-Bélanger, E. 2004. *Crisis Centres and Violence against Women: Dialogue in the Barents Region*. Oulu: Oulu University Press.

Skarhed, A. 2010. Evaluating the Swedish Ban on the Purchase of Sexual Services 1999–2008. SOU. Access at: www.government.se/contentassets/3df7f4e9fd1f44d58ecaca06f4045b56/press-releases-20062010—beatrice-ask.

Skilbrei, M.L. and Holmstrom, C. 2016. *Prostitution Policy in the Nordic Region: Ambiguous Sympathies*. Farnham: Taylor and Francis.

Sloan, L. 2006. *Women and Natural Resource Management in the Rural North: Arctic Council Sustainable Development Working Group 2004–2006*. Kvinneuniversitetet Nord: Forlaget Nora.

Stoor, J.P.A., Kaiser, N., Jacobsson, L., Salander Renberg, E. and Silviken, A. 2015. 'We Are like Lemmings': Making Sense of the Cultural Meaning(s) of Suicide among the Indigenous Sami in Sweden. *International Journal of Circumpolar Health* 74.

Stuvøy, K. 2014. Human Security and Women's Security Reality in Northwest Russia. *Environmental and Human Security in the Arctic*, 231–249.

Sumarokov, Y.A., Brenn, T., Kudryavtsev, A.V. and Nilssen, O. 2014. Suicides in the Indigenous and Non-Indigenous Populations in the Nenets Autonomous Okrug, Northwestern Russia, and Associated Socio-Demographic Characteristics. *International Journal of Circumpolar Health* 73(1): 24308.

Swedish Ministry of Justice. 2006. *Final Report of the Task Force Against Trafficking in Human Beings in the Barents EuroArctic Region*. Kajaani, Finland. www.barentsinfo.fi/beac/docs/11843_doc_TF-THBFinalReport.pdf.

UAA Justice Center. 2015. *Alaska Victimization Survey: Research on Violence Against Women in Alaska*. Council on Domestic Violence and Sexual Assault. www.uaa.alaska.edu/academics/college-of-health/departments/justice-center/research/alaska-victimization-survey/.

UNWomen. 2013. *Violence Against Women. Prevalence Data: Surveys by Country. Compiled by UN Women*. UN Women.

United Nations, and Social and Cultural Rights Committee on Economic. 2008. *Human Rights Instruments: Compilation of General Comments and General Recommendations Adopted by Human Rights Treaty Bodies*. Geneva: United Nations.

United Nations Development Programme. 1994. *Human Development Report 1994*. Oxford: Oxford University Press.

United Nations Development Programme. 2007. *Human Development Report 2007: Climate Change and Human Development*. 2007. Basingstoke: Palgrave Macmillan.

United Nations Development Programme. 2011. *Human Development Report 2011: Sustainability and Equity: A Better Future for All*. New York; Basingstoke: United Nations Development Program; Palgrave Macmillan.

United Nations General Assembly, Canada and Human Rights Directorate. 1985. *International Convention on the Elimination of All Forms of Racial Discrimination*. Ottawa: Human Rights Directorate, Canadian Heritage.

United Nations, Human Rights Committee and Russia. 2015. *Concluding Observations on the Seventh Periodic Report of the Russian Federation*. CCPR/C/RUS/CO/7, 28 April.

Yeasmin, N. 2013. Implementation of the EU Immigration Policy in the Barents Euro-Arctic Region: Towards a Framework for Policy Analysis. *The Yearbook of Polar Law Online* 5(1): 603–640.

Yeasmin, N. 2016. Theoretical and Empirical Justification for Anti-Immigrant Rhetoric: An Analysis on Diversity and Development from a Finnish Perspective. *Futura* 35(4): 26–32.

Young, T. K, Revich, B. and Soininen, L. 2015. Suicide in Circumpolar Regions: An Introduction and Overview. *International Journal of Circumpolar Health* 74(1).

2.8 Community security of indigenous peoples in the Barents Region

Giuseppe Amatulli and Joëlle Klein

This chapter addresses community security in the context of the Barents Region. Community security is one indicator that forms the broad concept of human security, and is closely interrelated with other aspects of human security, including environmental, political, economic, health and personal security. Given the geographic significance of indigenous groups in the northernmost part of the Barents Region, and the unique threats and opportunities that existing global challenges such as globalisation and climate change pose for northern communities, this chapter will focus predominantly on community security as it relates to indigenous Sámi, Nenets and Veps peoples.

1 Definition

Community security aims to address threats to groups of people that result in the loss of collective relationships and values, or exacerbate tensions between communities that lead to sectarian and ethnic violence. The UNDP defines community security as 'the capacity of people to derive their security through being active members of a group' (UNDP 2009, 13). The collective belonging that comes from participation and identification within a community can serve as a source of values and traditions to support individual wellbeing, and is thereby vital to the promotion of cultural identity (UNDP 1994, 31). Moreover, it is useful to underline that the term 'community' does not refer only to the individuals that form it, but also to other actors (such as civil society organisations and local authorities) that may play a key role in ensuring the vitality of a specific community living in a certain area (UNDP 2009, 15–16).

In this context, community security is directly linked to the realisation of human rights, which seek to ensure states and international actors meet their obligations to respect, protect and preserve the rights of individuals and communities. Furthermore, in communities where ethnic violence, terrorism or militarised armed force are not present, community security focuses on the social cohesion and positionality of marginalised or vulnerable groups within a group, or the positionality of a community in the dominant governance and social structures. Under Human Rights law, minorities are guaranteed a form of community security through Article 27 of the International Covenant on Civil and Political Rights (ICCPR). At the time of its drafting, indigenous peoples were considered minorities under this framework. Therefore, the provisions outlined in Article 27 are also applicable to indigenous peoples. Article 27 of the ICCPR stipulates:

> In those States in which ethnic, religious or linguistic minorities exist, persons belonging to such minorities shall not be denied the right, in community with the other members of their group, to enjoy their own culture, to profess and practice their own religion, or to use their own language.
>
> (UN General Assembly 1966)

The provisions enshrined in the above-mentioned article clearly indicate that states should not interfere with the rights that minorities are entitled to enjoy. According to the General Comment No. 31 issued by the UN Human Rights Committee (UNHRC), the treaty monitoring body under the ICCPR, negative obligations require states to abstain from violating the rights enshrined in the Convention. In addition, states must prove the need for enacting any restrictions on the enjoyment of certain rights, and can only take such actions in the pursuit of legitimate aims (UNHRC 2004). Nevertheless, both positive and negative obligations must be considered. To this degree, General Comment No. 23 issued by the UNHRC has provided a comprehensive explanation of the provisions of Article 27. General Comment No. 23 establishes that negatives states' obligations are not sufficient to guarantee the individual and collective rights of minorities, and therefore require positive actions on the part of the states. Specifically, in order to protect and ensure the rights of a certain community, positive legal measures are required, with the aim of guaranteeing the real and effective participation of community members in decision-making on issues that impact them (UNHRC 1994, para. 6.2, 7). The positive obligations that a state must implement to ensure the enjoyment of rights enshrined in the ICCPR are not limited to the actions of states' agents, but also include protection against violations that private persons or an entity can commit. The legal obligations concerning the individual and collective rights of indigenous peoples have also been specified under ILO Convention 169 (ILO 1989), a treaty that has only been ratified by Norway in the Barents countries, and the UN Declaration on the Rights of Indigenous Peoples adopted by an overwhelming vote of 163 states at the UN General Assembly in 2007, but for which Russia has abstained to vote.

However, community security is not exclusively defined through legal collective rights, and requires definition through the framework of human security. The principles of human security necessitate a people-centered approach, and therefore the definition of community security in relation to indigenous populations must also inherently be self-articulated (UNTFHS 2016, 6). In stakeholder reports to the UN Human Rights Council presented by the Sámi Parliament and other indigenous groups, issues pertaining to language preservation, education in native languages and the impact of human activities on livelihood are articulated as issues vital to community security (HRC 2015, 8–10; HRC 2013, 10–11; HRC 2014, 8–10; HRC 2017, 6–8). Furthermore, the regional community security goals articulated by the Working Group on Indigenous Peoples in the Barents Euro-Arctic Region (hereafter WGIP) specify 'The overall goal for the indigenous peoples' cooperation in the Barents Euro-Arctic Region is to secure indigenous peoples' rights, foundation for trade, society, culture and language … [and secure] solid health- and living conditions, as well as cultural continuity' (WGIP 2017, 7). Therefore, a regional and community-based understanding of the protection of community security in the Barents includes ensuring that the development and preservation of collective indigenous identity, livelihood, language and culture is self-determined, protected within national borders or in transnational contexts, and respected by other actors and authorities.

Furthermore, other minorities identified by gender, sexual orientation, (dis) ability, age and class exist within (and are important aspects of) the composition and security of indigenous communities. The broad scope reflected in defining community security therefore requires a narrowing in order to assess aspects of security unique to the Barents Region. In this regard, community security can be understood to include both an 'external component', and an 'internal component', which differentiates cooperation and security in the relationships between indigenous communities and other minorities, from the relationships between indigenous communities and local, regional and national state authorities (WGIP 2017, 6). The external component is linked with the protection of the community's ability to develop and promote its culture and traditions, and maintain and protect its collective rights. This is connected to the concept of self-determination, which seeks to preserve the autonomy of a community to participate in decision-making on issues that affect them. The internal component is linked to social cohesion within and between communities, addressing the presence of minorities within minorities, and the individual pressures and threats faced by community members by their own or co-existing communities. The latter component intersects with the concept of personal security, as discussed in Chapter 2.7, and additionally, analysis on the security of double minorities in the Barents Region would require a more in-depth and comprehensive study beyond the scope of this book. Therefore, this chapter will focus on the 'external components' to analyse and discuss threats to indigenous community security in the Barents Region.

2 Contextualisation

The geographic boundaries of the Barents Region predominantly include the homelands of indigenous peoples. While diverse communities inhabit the Barents Region, only three are formally recognised as indigenous peoples by their respective national governments. In the Nordic countries, the Sámi are recognised within the national framework of Finland, Norway and Sweden. In the Russian Federation, only the Nenets, Veps and Sámi are recognised as indigenous peoples, notwithstanding the presence of other indigenous communities living the Russian Federation (such as the Komi and Pomors) (Barentsinfo 2017).

An assessment of the regional conditions of indigenous peoples living in the Barents needs to account for the considerable dissimilarities between the Northwest part of Russia and the Nordic countries. Such differences include the unique linguistic, cultural, political, economic and livelihood traditions amongst minorities and indigenous groups inhabiting the region (Zimmerbauer 2013, 94). Furthermore, the federal legal definition of indigenous peoples in Russia differs from the definition utilised in the Nordic countries. In Russia, indigenous peoples are recognised in part by their population size; groups with less than 50000 people are defined as 'indigenous numerically small peoples,' and non-Russian groups with more than 50000 peoples are not able to claim such status (Barentsinfo 2017).

Indigenous representation in the Nordic countries includes more formalised status in political institutions, *Sámediggis* (Sámi Parliaments), while Russian indigenous representation does not include formal political organisations, and any advocacy must be conducted through non-governmental organisations (Mörkenstam 2016, 9–13; Vinding and Mikkelsen 2016, 43–44; Rohr 2014, 9) This can be problematic for the protection of collective and individual human rights, as indigenous non-profit organisations in Russia have been directly threatened by 'foreign agent' laws requiring additional registration and documentation procedures if they receive funding from a foreign source (HRW 2017). Courts can leverage hefty fines on non-profits if they are unable to comply, which has resulted in the closure or restricted activity of many organisations including ten civil society and indigenous organisations in the Russian Barents Region (The Independent Barents Observer 2017; HRW 2017).

Many indigenous populations in Russia do not receive the federal status of 'indigenous numerically small peoples,' which limits the protections of indigenous rights for such communities. Indeed, without recognition, they cannot establish local democratic self-governance to protect and promote their specific interests and rights. Moreover, local governance fosters increased autonomy, and provides a more legitimate basis for inhabitants to control community development, which increases the sense of belonging amongst individuals in a specific community (Tennberg 2012, 135–136).

Komi peoples are not recognised as indigenous peoples within the Russian national framework, although the legislation of the Komi Republic grants them the status of indigenous peoples (Barentsinfo 2017). The Komi live across the Komi Republic, the Murmansk Oblak, the Arkhangelsk Oblak and the Nenets Autonomous Okrug. Komi peoples have historically faced discrimination based on their identity, which, in conjunction with policies from the Soviet era, has led to unequal education achievement levels between Komi and Russians in the Komi Republic (Barentsinfo 2017). Other minorities including the Pomors and Karelians are not recognised as indigenous groups in the Barents, but face similar ethnic discrimination.

Veps peoples are recognised as indigenous people in the Russian legislation as of March 2000. The majority of Veps live in remote villages in the south of the Republic of Karelia and in the Volodga and Leningrad Oblasts. According to the last census before the collapse of the Soviet Union, the number of Veps has decreased from 12000 Veps in 1989 to less than 6000 in 2010 (Barentsinfo 2017). In this regard, it is important to note that in twenty years (from 1989 to 2010), the Veps have witnessed a population decline of about 50%. Furthermore, according to the 2002 census, only 37.5% considered Veps as their native language (Barentsinfo 2017). The Veps language is considered a vital component of the promotion of their cultural identity, and to this degree, the decreasing population in conjunction with a reduced number of native speakers underlines the insecurity surrounding their cultural preservation in the long-term.

The Nenets inhabit diverse areas of northern Russia, but are predominantly located in the Komi Republic, the Nenets Autonomous Okrug, Arkhangelsk Oblast and Murmansk Oblast. According to the 2010 census, Nenets people constitute the biggest indigenous group living in Russia, with a population of around 44640 people (Rohr 2014). Although this represents a general increase since 2002, Nenets inhabiting the European tundra in the Barents Region have continued to experience population decline and out-migration (Nymand Larsen 2014, 56). They also face considerable community challenges, such as the disappearance of the Nenets language, unemployment and inadequate access to healthcare and services (Barentsinfo 2017). Furthermore, suicide and low life expectancies are also prevalent, and the mortality rate among Nenets is almost four times higher in comparison to the average in Russia (JS 3 2012, 8; Barentsinfo 2017). In terms of economic opportunities and livelihood, unemployment remains high amongst indigenous peoples of the Yamal Nenets Okrug at 24.5%, and the incomes of indigenous peoples in Russia are well below the Russian national average (JS 3 2012, 8; Barentsinfo 2017).

For Sámi and Nenets living in the Barents, traditional livelihoods such as reindeer herding and fishing practices are a vital component of their cultural and indigenous identity (Kaiser et al. 2013, 3; WGIP 2017, 10). Generally, collective reindeer herding (which implies cooperation in relation to collective ownership) is the main form of herding practised by in the Barents, and is inherently trans-boundary in nature. Such practices are a predominant source of income for many indigenous peoples in the Barents; in the last decades, reindeer herders have started to produce and sell reindeer meat and products in the global market, and demand has increased the potential to derive income from such business, especially for the Russian Barents (WGIP 2017, 10). Notwithstanding such developments, insecurities still impact herding livelihoods, as herding is disrupted by changing weather patterns and other human activities such as mining. In addition to this, the confluence of natural resource use and exploitation with the out-migration and dwindling interest of young generations in traditional lifestyles in favour of more economically stable employment poses a serious risk for the survival of reindeer herding practices in the area (Barentsinfo 2017).

The Sámi are the only recognised indigenous people in the European Union and their homeland, Sápmi, is located in the northernmost part of Norway, Sweden, Finland and the Russian Kola Peninsula. Approximately 50000–65000 Sámi live in Norway: 20000 in Sweden; around 8000 in Finland and almost 2000 in the Peninsula of Kola (Vinding and Mikkelsen 2016, 29). The Sámi are recognised as indigenous peoples in the Constitution of Finland, peoples in Sweden, and have been granted specific rights in the Norwegian Constitution. Concerning the Russian national framework, the Sámi are recognised under the status of 'indigenous numerically small peoples'. In each Nordic country, there are different statutory assemblies to represent the Sámi, and these parliaments or *Sámediggis*, additionally appoint representatives to the trans-boundary Sámi Parliamentary Council (SPC). The SPC is the co-operational body for the Finnish, Swedish and Norwegian Sámi Parliaments. It is open also to Russian Sámi organisations, but given the lack of an elected body representing the Sámi in the Russian Federation, two Russian Sámi organisations have instead been made permanent representatives in the SPC (Sámediggi, 2017a). In the Russian Federation, the Sámi are represented by the Russian Association of Indigenous People of the North (RAIPON), which has the same features of the Sámi Parliamentary Council.

Regarding the positionality of Sámi rights in the international arena, only Norway has ratified the ILO Convention No. 169 (Convention concerning Indigenous and Tribal Peoples in Independent Countries), while Sweden and Finland continue to address the issue of ratification on the parliamentary level. Despite this, all the three Nordic countries voted in favour of the adoption of the UN Declaration on the Rights of Indigenous Peoples (hereafter UNDRIP) in 2007. Moreover, all Barents countries have ratified the ICCPR; therefore, Sámi people at a minimum receive protections articulated for minorities, under article 27 of the Convention. Furthermore, the governments of the Nordic states are in the process of adopting the Nordic Sámi Convention, which would strengthen and enhance the recognised rights of Sámi people, especially in a cross-border cooperation context in the northernmost areas of Finland, Norway and Sweden (Cambou 2018, Koivurova 2008). One of the main goals of the Convention is to address challenges regarding Sámi rights in the context of providing a comprehensive framework to regulate the legal relationships between the Sámi and the three Nordic countries (Koivurova 2008, 104). Among the objectives of the Convention, there is a duty to enable Sámi people to exercise the right to self-determination, safeguard and develop their languages, culture and traditional livelihoods; respect Sámi people's conceptions of legal traditions, customs and law. Important provisions regarding language preservation, such as the right to produce media and receive education in Sámi language, and safeguards for Sámi rights to land and water, and the protection of Sámi traditional livelihoods, are included in the Convention (Draft Nordic Sámi Treaty 2017).

The international positionality of the Sámi and other indigenous peoples in the Russian Federation, however, remains much more limited. Article 27 of the ICCPR remains the only provision to which indigenous peoples living in Russia can appeal to for a recognition of their rights and, as explained above, was primarily conceived to protect minorities. There are no other provisions that indigenous peoples of Russia can appeal to, given that the Federation did not ratify ILO Convention No. 169, abstained during the vote for the adoption of the UNDRIP in 2007, and is not involved in the draft of the Nordic Sámi Convention. In the particular case of the draft Nordic Sámi Convention, the Expert Committee strongly supported that the three Nordic countries cooperate with Russia in order to include the Russian Sámi (Cambou 2018, Koivurova 2008, 109). This can be seen as an attempt to extend the validity of the Convention to the Russian Sámi, but without involving the Russian State into the negotiations. The Committee's hesitance to involve Russia in the process is grounded both in Russia's abstention from the adoption of the UNDRIP, as well as the diversity of indigenous peoples present in Russia; it would have been difficult to take part in an international convention for Russian Sámi, while leaving other indigenous groups without any specific protections (Koivurova 2008, 108–109).

As previously discussed, indigenous community security is broadly articulated in the human rights framework at the international level through the UNDRIP, ILO C169, the draft Nordic Sámi Convention and other human rights documents. However, in order to address aspects of indigenous community security specific to the Barents Region, further contextualisation within regional structures is necessary. The Working Group of Indigenous People (WGIP), which holds advisory status to the Barents Euro-Arctic Council (BEAC) and the Barents Regional Council (BRC), is composed of representatives from the national indigenous populations of each Barents states. In the Nordic countries, representatives are appointed by the Sámi Parliamentary Council, while representatives of the Sámi, Nenets and Veps in Russia are appointed by the Sámi Congress, Yasavey and Vepsian Cultural Society, respectively (WGIP 2017, 4). Together, the WGIP developed the 2016–2018 Action Plan for Indigenous Peoples in the Barents Euro-Arctic Region (BEAR). The Action Plan defines current priorities in the development of indigenous peoples' communities and societies in the region, and identifies areas for cooperation between indigenous communities and states to promote such interests (WGIP 2017, 3). Although the legal significance of the Action Plan is minimal, the document is a regional and self-articulated representation of the threats and priorities present in indigenous communities in the Barents. It is used in this chapter to guide the assessment of indigenous community security in the Barents Region, as it serves to reflect indigenous voices on community security, and preserve the principle of a people-centred approach.

3 Assessment

Historically, indigenous peoples in the Barents have faced discrimination and inequality through the imposition of state assimilation processes, and the impacts and effects are still visible in the insecurity of communities today. Cultural preservation through education, knowledge sharing and language are still complex and insecure for indigenous communities in modern contexts. Furthermore, environmental pressures triggered by human activities and resulting from climate change also threaten the traditional land-use and environmental practices of indigenous peoples in the region. Ultimately, securing the adequate participation of indigenous peoples in decision-making and legislation is fundamental to articulating indigenous rights and achieving community security. Although communities have identified more specific priorities in the realm of economic, environmental and personal security, these elements are discussed more broadly in other chapters.

3.1 Cultural preservation as the core of indigenous community security

In the Action Plan, it is recognised that the indigenous peoples of the Barents' (the Nenets, Sami and Veps) 'way of living, [their] culture, and languages have historical roots in the area, and future development must take this into account' (WGIP 2017, 6). In the past century, indigenous communities in the Barents have faced discrimination and cultural assimilation, through the enforcement of harsh state assimilation policies. In Norway, Norwegianisation became an official policy at the end of the 19th century, and sought to assimilate Sámi into Norwegian society through the erasure of non-Norwegian cultural practices and identification (Joona 2012, 58; Minde 2005, 6). In addition to this, after the Second World War, the compulsory education and the affirmation of the welfare state had a significant role in the assimilation of Sámi people within the Norwegian state (Cambou 2016, 300–301). The impact of these policies increased formal and informal discrimination against the Sámi, and drastically reduced the number of people willing to identify as Sámi, or able to speak a Sámi language. Such policies were prevalent across the Barents, with similar policies in Sweden, Finland and the Soviet Union attempting to integrate Sámi, Nenets and Veps into the national population and dominant culture (Minde 2005, 11-ss.).

In recognition of the impact that assimilation policies had on the existing culture and identity of Sámi, there was a shift in the national public policy in addressing Sámi issues in the 1960s. Additionally, the inclusion of indigenous peoples in the international arena in the late 1990s has resulted in increased awareness, advocacy and rights protections for indigenous communities. To this degree, the history of cultural assimilation has led to a prioritisation of the right to cultural preservation as the key element underlying indigenous community security. As the Action Plan denotes, '[Barents indigenous peoples'] cultures are closely attached to primary industries, the nature, their languages, traditions and history through solidarity, socialization and identity' (WGIP 2017, 13). However, despite achievements Regarding the realisation of civil, economic and social rights of indigenous communities in the Barents are not comprehensive, as indigenous identities were still consistently defined by state actors with the influence of prejudices and stigmatisations that were present in the mainstream society (Pikkarainen and Brodin 2008, 21).

Furthermore, the preservation of indigenous culture through language and education practices is a priority and challenge for communities in the Barents Region. As is noted in the Action Plan '[e]ducation is the key to transmission of indigenous languages...[a] satisfying level on indigenous peoples' educational institutions is a central point for further development of the competence level among indigenous peoples' (WGIP 2017, 11). Specifically, the realisation of the rights of indigenous peoples to receive education in their native language is a serious problem (WGIP 2012, 12). Access and availability of resources to support education and teaching programs in native languages remain the greatest barrier to the realisation of such rights (Anaya 2012, 21–22). Indeed, local authorities have expressed that often no suitable teachers are available to provide quality education in indigenous languages, due to a lack of available and qualified candidates (Anaya 2012, 19; Council of Europe 2017).

3.2 Indigenous language and education

In the Barents Region, the preservation of indigenous languages varies between Russia and the Nordic countries. In the Nordic countries, several Sámi languages exist; the most commonly spoken are Inari Sámi, Skolt Sámi and North Sámi (Sámediggi, 2017b). According to the provisions enshrined in the Sámi language Acts of the three Nordic states, Sámi peoples are entitled to receive education in a Sámi language at school and, moreover, they are entitled to address the public administration and to go to court using their native language (Chelstowska 2009, 19). Despite this, Inari and Skolt Sámi languages in conjunction with smaller Sami languages such as Lule and Southern Sámi are still classified as languages in danger of extinction (Anaya 2012, 19). Furthermore, an additional challenge is the development of curriculum in state funded schools that includes and incorporates Sámi history and culture into mainstream education (Anaya 2012, 18–19).

The situation regarding indigenous rights to language preservation is extremely delicate in the Russian part of the Barents, where assimilation policies and the lack of implemented legislation on language preservation have been complicit in the extinction and slow disappearance of languages. The majority of Russian Sámi people live in the Murmansk Oblast and the Kola Peninsula where Skolt Sámi, Ter Sámi, Kildin Sámi and Akkala Sámi are the most common indigenous languages. Among others, Akkala and Ter Sámi languages have almost disappeared; indeed, Ter Sámi is now only spoken by two people worldwide. This is the result of a general lack of proper policy, and Russia's unwillingness to ratify international documents on the protection of indigenous languages. While all three Nordic countries have ratified the European Charter for Regional or Minority Languages, Russia has not. In addition, the Russian Constitution includes only vague provisions on the necessity to guarantee linguistic rights; but to date, no specific language acts have been conceived to properly ensure the linguistic rights of indigenous and minority groups living in the Russian Federation (Chelstowska 2009, 21).

Although the Sámi, Nenets, Veps and Komi languages have no official recognition, there has been a significant shift in public services in the last decades, especially in comparison to harsh assimilatory policies of the Soviet era. In the past, curriculum was available only in Russian, while today there are a limited number of schools in which Sámi, Nenets and Komi are the primary language of instruction until the fourth grade (Barentsinfo 2017). Following the collapse of the Soviet Union in 1989, the Veps Cultural Society was established, with a primary aim to increase knowledge of the Veps language (Kolga 2001). Their main activities included the publication of books in Veps, which are used in schools offering Veps curricula in the Republic of Karelia. Furthermore, radio broadcasts in Sámi, Komi and Veps languages exist, in addition to few newspapers and magazines in Kildin Sámi, Nenets, Komi and Veps languages (Barentsinfo 2017). However, general educational materials contain limited information about indigenous culture and history and, in addition to this, the language is seldom used at home (Chelstowska 2009, 25). The primary use of national languages at home and at school, in addition to a lack of adequate educational materials concerning indigenous history and culture, is a prevalent concern across all indigenous communities in the Barents Region. To this degree, it can be recognised that many languages vanished or are vanishing due to national policies that do not take into consideration the linguistic rights of indigenous peoples (Pikkarainen and Brodin 2008, 24–25).

Furthermore, indigenous peoples who specifically claim their linguistic rights are often seen with prejudice by the authorities. In particular, indigenous peoples face discrimination and prejudice within healthcare systems; which are often related with their linguistic rights, or due to inaccessible or a lack of culturally appropriate services (CERD 2013, 29–30; CERD 2015, 27–28). In the Nordic Sápmi, parents express having experienced discrimination through their interactions with doctors, nurses and therapists, in regards to their children utilising a Sámi language in order to communicate with family members (Pikkarainen and Brodin 2008, 30). Furthermore, studies have found that children face teasing and social discrimination because they speak a Sámi language as their mother tongue, instead of the dominant national language (Pikkarainen and Brodin 2008). Additionally, elderly indigenous peoples also face linguistic discrimination, which, in conjunction with their histories of trauma from past assimilation policies, create a nuanced need for culturally and linguistically appropriate services. Municipalities in which Sámi are present are required to offer eldercare services (entirely or partially) in a Sámi language, as there are still communities of elderly Sámi that exclusively utilise a Sámi language to communicate with others. As in education, the greatest barrier to meeting such needs is the availability of qualified candidates and employees, but additionally municipalities lack the political will to protect the use of Sámi languages within the Sámi eldercare sector (Pikkarainen and Brodin 2008, 31).

Indigenous peoples of the Barents also face discrimination in labour and employment sectors. On different occasions, notwithstanding university degrees in social or natural sciences for example, indigenous peoples have been advised to look for a job in the tourism sector, due to their ethnic origin (Pikkarainen and Brodin 2008, 31–32). In addition to the perceived commercialisation of their culture and essentialisation of their value to society as a subject for touristic consumption, indigenous people experience a social environment that does not seriously consider their competences, resulting in a perpetuation of negative perceptions regarding the appropriateness of certain types of work for indigenous communities (Pikkarainen and Brodin 2008, 32). In conclusion, indigenous peoples living in the Barents continue to experience linguistic and ethnic discrimination, and by extension, the lack of appropriate protections for education and linguistic rights continue to exist as threats to community security.

3.3 Land and natural resource rights

Discrimination against indigenous peoples in the Barents and the trauma of assimilation processes are also connected to the right to utilise traditional land and resources. Indigenous community security in the Barents Region is inherently linked to land and natural resource rights, 'the economic development of the region must increase the participation of the indigenous peoples in the fields of protection and development of their own communities…[*sic*] [and is] decisive for the survival of the indigenous peoples' communities' (WGIP 2017, 10). Historically, indigenous people have been excluded from any kind of participation in decision-making, which has led to contemporary disputes with states and companies seeking to govern or utilise the existing natural resources present on traditional lands. Indeed,

> The fundamental root source of conflict between indigenous peoples, on the one hand, and States and non-indigenous entities and individuals, on the other, is their differing views as to which actor possesses valid title to the land and resources located in territories traditionally occupied by indigenous groups.
>
> (Economic and Social Council 2004, 5)

Currently, resource related conflicts between states, indigenous groups, and non-indigenous groups are an influential factor in shaping the local politics of natural resource management and land rights in the Barents Region. For example, in Finland recent fishing quotas and licensing regulations agreed upon by Finland and Norway around the Teno River region have spurned protests by Sámi peoples and fishermen (YLE 2017). In Norway, protests against mining activities and wind farms have increased, as Sámi reindeer herders and fishermen are increasingly concerned about the impacts of such activities on herding districts and water quality (Vinding and Mikkelsen 2016, 32). Similarly, in Sweden, Sámi rights to traditional fishing and hunting practices have been challenged, and state permission and support for mining companies have raised concerns among Sámi activists over the protection of their traditional lands and practices (Vinding and Mikkelsen 2016, 34). In Russia, amendments to the existing land codes have enabled diamond and gold mining activities to increase, without a meaningful attempt to acquire the consent of indigenous peoples. Furthermore, oil and liquid natural gas (LNG) projects threaten traditional fishing practices, and the reindeer herding livelihoods of Nenets on the Yamal Peninsula, which further underscore the lack of indigenous rights provided and protected in the Russian legal framework (Vinding and Mikkelsen 2016, 47–48; Rohr 2014).

According to the provisions established in the UNDRIP, indigenous peoples are entitled to maintain traditional practices on occupied lands and waters (article 25); they have the right to use, develop and control the lands, territories and resources that they have been using since time immemorial (article 26), and they are entitled to determine and develop specific strategies for the development of their lands (article 32). All the activities that take place on indigenous lands and impact indigenous traditional lifestyles and livelihoods require the Free, Prior and Informed Consent (hereafter FPIC) of the relevant indigenous group. As established in the Convention, the need for FPIC from indigenous groups is established in several cases, including: relocation (article 10); undue expropriation of their lands and resources (article 11–28); before adopting any legislative measure that can impact them (article 19); before the approval of any project that can impact their lands (article 32), and in cases where hazardous materials are to be deposited on indigenous lands (article 29). The importance of these rights is central to the community security of indigenous groups in the Barents, and articulated in the Action Plan as '[the] transfer of knowledge and competence regarding political work and international legislation regarding indigenous peoples' rights is a priority of the WGIP during the Action Plan period' (WGIP 2017, 13).

In the Barents Region, indigenous peoples' enjoyment of land and natural resources rights depends on the existing national legislation in each of the Barents states. Until the 1960s, it was generally accepted that at the moment of annexation, states automatically gained possession of 'ownerless lands'. States justified their sovereignty over these lands under the principle of 'terra nullis', which applies to territories that had never been subject to any kind of sovereignty, or where previous sovereigns had, implicitly or explicitly, forsaken sovereignty (Joona 2012, 66; Gilbert 2006). This was problematic, as it negated the existence and agency of indigenous peoples inhabiting such territories and therefore resulted in the governance of such lands by sovereign states without indigenous consent, consultation or participation. In the contemporary political sphere, this would amount to a violation of the right to self-determination, and disregard for FPIC in dealing with indigenous rights to land and resources.

The main institutions conceived to enhance the self-determination of Sámi peoples in the Barents are the three Sámi Parliaments of Norway, Finland and Sweden. Although this does not overtly include Russian indigenous groups, these institutions serve as important models in building and strengthening indigenous self-governance and participation in decision-making across the Barents Region, especially on local cultural and governance issues. However, their level of involvement in decision-making and their real power to influence policymakers at the State level is still limited and differs between the different national legislations (see also Chapter 2.9 on political security).

Specifically, in Finland the Sámi Parliament is limited to decision-making that concerns Sámi culture and language; while it is not involved in matters related with the use of natural resources and the right to land (Anaya 2012, 11). Nevertheless, the Finnish government is working on the ratification of ILO C169 and Finland does officially recognise Sámi as indigenous people in the Finnish Constitution (Constitution of Finland 1999, Chapter 2, section 17), and many national laws (like the Finnish Mining Act and the Finnish Reindeer Husbandry Act) to ensure that their rights are protected and safeguarded. The same situation is present in Sweden, where the Sámi Parliament is involved in minor matters, and the Swedish Government is considering assessments on the possibility of ratifying ILO C169. Nevertheless, in national legislation, Sámi rights are safeguarded through specific Acts like the Swedish Reindeer Herding Act and the Swedish Minerals' Act, and such laws still lack specific provisions on the inclusion of Sámi people in consultation processes. On the contrary, Norway remains the only Barents state that has ratified ILO C169, and more recently signed a consultation agreement with the Norwegian Sámi Parliament. Notwithstanding the importance of such an agreement, such consultation does not necessarily guarantee the inclusion of Sámi in all decision-making. Indeed, the Sámi Parliaments have reported that on many occasions, governments have failed to start consultations prior to making and implementing decisions, and begun consultations after decisions have already been made (Anaya 2012, 11–12). The situation is more difficult for the Sámi, Nenets and Veps peoples living in Russia, given that the Federation has shown little possibilities to ratify the ILO C169, or to adopt specific provisions in order to safeguard the rights to land and natural resources of those indigenous peoples living within its national boundaries. The Russian Federation's tendency to ignore the rights of indigenous peoples was reconfirmed during the vote to adopt the UNDRIP, when Russia abstained (Koivurova 2008, 108–109).

The historic and unique relationship that indigenous peoples have to their lands unites community wellbeing to the welfare of the land and environment. As a result, the exploitation of natural resources directly impacts the ability of a community to determine the future of its culture (WGIP 2012, 11–12). In particular, land-based activities (such as reindeer-herding) form an integral part of indigenous peoples' traditional livelihoods, and as discussed briefly above, has led to conflicts with other stakeholders on many occasions (WGIP 2012, 5–7). The lack of indigenous involvement in decision-making, especially in cases of trans-national corporations carrying out exploitation processes in indigenous areas, has resulted in the need to abandon considerable portions of land in the Nenets tundra, or to the loss of lands that Sámi and Komi peoples used for reindeer herding or fishing in the Kola Peninsula (WGIP 2012, 13). To this degree, the rights of indigenous peoples in the Barents Region still require critical attention. Indigenous participation and support for self-determination in decision-making processes, especially concerning land and natural resource rights, need to be further enhanced. Within the BEAC, the Action Plan promotes 'Veps, Nenets, and Sami representation on all levels in decision-making processes' (WGIP 2017, 13). Ultimately, an increased recognition of indigenous rights at national and international levels is necessary, and community security relies on the positionality and representation of indigenous groups in decision-making processes.

4 Conclusions and recommendations

Ensuring community security for indigenous peoples living in the Barents Region is significantly dependent on their ability to participate in decision-making processes, and the articulation of indigenous security in the Barents needs to be people-centred. To this degree, it is clear that the foundation of indigenous community security rests on indigenous agency in cultural preservation, which can be achieved through a realisation of linguistic and educational rights, in addition to land and natural resources rights.

In order to mitigate the negative environmental and community impacts that industrial activities poses on the Barents Region, a substantial improvement in the existing legislation is needed. Indigenous community security in the Barents Region rests on a need for comprehensive cooperation across indigenous communities, locals and policy makers. In this sense, cooperation requires inclusion in multiple aspects: inclusion in public life, participation in public debate and inclusion in decision-making processes. Inclusion is vital to security as people both 'derive their security through being active members of a group' and '[develop] active and inclusive relationships between residents, private sector, public sector and civil society organizations, in order to foster individual and collective well-being' (UNDP 2009, 13; Berger-Schmitt 2000, 406). The ultimate aim of inclusion in community security thus is to 'develop a community of shared values, shared challenges and equal opportunity' (Scott 2010, 56). Hence, indigenous community security in the Barents Region is ensured in a human rights context through two main aims: involving peoples in decision-making, and improving legal existing tools. In light of this, ensuring the inclusion of indigenous peoples in decision-making in alignment with existing tools such as ILO 169 and the UNDRIP serves as the basis for ensuring indigenous community security in the Barents Region.

References

AnayaJ. 2012. *Report of the Special Rapporteur on the rights of indigenous peoples.* UN Doc. A/HRC/18/35. http://unsr.jamesanaya.org/docs/countries/2011-report-sapmi-a-hrc-18-35-add2_en.pdf.

Barentsinfo. 2017. General information portal to the Barents Region. www.barentsinfo.org/Barents-region

Berger-SchmittR. 2000. *Social cohesion as an aspect of the quality of society: concept and measurement.* Mannheim: Centre for Survey Research and Methodology.

CambouD. 2016. The legal significance of the right of indigenous peoples to self-determination and its implications for the Sámi people. PhD thesis. Vrije Universiteit Brussels.

CambouD. 2018. The 2005 Draft Nordic Sámi Convention and the Implementation of the Right of the Sámi People to Self-determination. In *Critical indigenous rights studies: new directions in indigenous rights research,* edited by G. Corradi, K. De Feyter, E. Desmet and K. Vanhees. Abingdon: Routledge.

Committee on the Elimination of All Forms of Racial Discrimination. CERD. 2015. *Consideration of reports submitted by States parties under article 9 of the Convention.* Twenty-third periodic reports of States parties due in 2015 – Finland (23 December), CERD/C/FIN/23.

Committee on the Elimination of All Forms of Racial Discrimination. CERD. 2013. *Consideration of reports submitted by States parties under article 9 of the Convention.* Twenty-first and twenty-second periodic reports of States parties due in 2013 – Norway (6 September), CERD/C/NOR/21–22.

ChelstowskaN. 2009. Language policies in Northern Europe. The case of Sámi minorities in Scandinavia and Russia. MA Thesis (General Linguistics), University of Amsterdam.

Constitution of Finland. 1999. www.finlex.fi/en/laki/kaannokset/1999/en19990731.pdf.

Council of Europe. 2017. *Sixth report of the Committee of Experts in respect of Sweden, 10/5/2017.* https://search.coe.int/cm/Pages/result_details.aspx?ObjectID=0900001680703ca5#_Toc47597 0368

Draft Nordic Sámi Treaty. 2017. www.samediggi.fi/index.php?option=com_docman&task=cat_view&gid=191&Itemid=10

Economic and Social Council. 2004. *Annotations to the provisional agenda. Prepared by the Secretariat pursuant to the request of the Working Group on Indigenous Populations at its twenty-first session.* UN Doc.E/CN.4/Sub.2/AC.4/2004/2. www2.ohchr.org/english/issues/indigenous/docs/wgip22/1add1.pdf.

GilbertJ. 2006. *Indigenous peoples' land rights under international law: from victims to actors.* New York: Transnational Publishers. Human Rights Council. 2017. *Working Group on the Universal Periodic Review. Summary of stakeholders' submissions 1 on Finland – Report of the Office of the United Nations High Commissioner for Human Rights.* Twenty-seventh session – Finland (1–12 May), A/HRC/WG.6/27/FIN/3.

Human Rights Council. 2014. *Working Group on the Universal Periodic Review. Summary prepared by the Office of the United Nations High Commissioner for Human Rights in accordance with paragraph 15 (b) of the annex to Human Rights Council resolution 5/1 and paragraph 5 of the annex to Council resolution 16/21 – Norway.* Nineteenth session – Norway. (28 April–9 May), A/HRC/WG.6/19/NOR/3.

Human Rights Council. 2013. *Working Group on the Universal Periodic Review. Summary prepared by the Office of the High Commissioner for Human Rights in accordance with paragraph 5 of the annex to Human Rights Council resolution 16/21 – Russian Federation*. Sixteenth session – Russian Federation. (22 April–3 May), A/HRC/WG.6/16/RUS/3.

Human Rights Council. 2015. *Working Group on the Universal Periodic Review. Summary prepared by the Office of the United Nations High Commissioner for Human Rights in accordance with paragraph 15 (c) of the annex to Human Rights Council resolution 5/1 and paragraph 5 of the annex to Council resolution 16/21 – Sweden*. Twenty-first session – Sweden (19–30 January 2015), A/HRC/WG.6/21/SWE/3.

Human Rights Watch(HRW). 2017. Russia: government vs. rights groups. www.hrw.org/russia-government-against-rights-groups-battle-chronicle.

ILO. 1989. *Convention (No. 169) concerning Indigenous and Tribal Peoples in Independent Countries*.

JoonaT. 2012. no. 169 in a Nordic context with comparative analysis: an interdisciplinary approach. *Juridica Lapponica*. Rovaniemi: LUP Lapland University Press

JS 3, 2012. *Joint Submission by the Russian Association of Indigenous Peoples of the North (RAIPON), International Work Group for Indigenous Affairs (IWGIA), Institute for Ecology and Action Anthropology (INFOE). NGO Submission to the Human Rights Council – Universal Periodic Review Mechanism*. UN Universal Periodic Review 16th Session, 22 April – 3 May.

KaiserN., RuongT. and Salander RenbergE. 2013. Experiences of being a young male Sami reindeer herder: a qualitative study in perspective of mental health. *International Journal of Circumpolar Health* 72.

KoivurovaT. 2008. The draft for a Nordic Saami Convention. *European Yearbook of Minority Issues* 6: 103–136.

KolgaM. (ed.). 2001. *The Red Book of the Peoples of the Russian Empire*. Tallinn: NGO Red Book.

MindeH. 2005. *Assimilation of the Sami: implementation and consequences*. Aboriginal Policy Research Consortium International (APRCi), Paper 196, Kautokeino, Norway.

MörkenstamU., JosefsenE. and NilssonR. 2016. The Nordic Sámediggis and the limits of indigenous self-determination. *Gáldu Čálá – Journal of Indigenous Peoples Rights*.

Nymand LarsenJ. (ed.), 2014. *Arctic human development report: regional processes and global Linkages*. Copenhagen: Nordic Council of Ministers.

PikkarainenH., BrodinB. 2008. *Discrimination of the sami: the rights of the sami from a discrimination perspective*. Stockholm: DO:s rapportserie.

RohrJ. 2014. *Indigenous peoples in the Russian Federation (No. 18)*. International Work Group for Indigenous Affairs.

Sámediggi. 2017a. *The Sámi Parliamentary Council*. www.samediggi.fi/index.php?option=com_content&task=view&id=11&Itemid=65&lang=english

Sámediggi. 2017b. *The Sámi languages*. www.samediggi.fi/index.php?option=com_content&task=view&id=23&Itemid=112

ScottK. 2010. *Community Vitality*. Waterloo, ON: Canadian Index of Wellbeing and University of Waterloo.

TennbergM. (ed.). 2012. *Politics of development in the Barents Region*. Rovaniemi: Lapland University Press.

The Independent Barents Observer. 2017. Kola Eco Centre labeled as 'foreign agents'. https://thebarentsobserver.com/en/civil-society-and-media/2017/04/kola-eco-centre-labeled-foreign-agents.

UN General Assembly. 1966. *International Covenant on Civil and Political Rights (ICCPR), including Optional Protocol*. 999 UNTS 171.

UNGeneral Assembly. 2007. *Declaration on the Rights of Indigenous Peoples.* UN doc. A/RES/61/295.

UN Human Rights Committee (HRC). 1994. *CCPR General Comment No. 23: Article 27 (Rights of Minorities).* UN Doc. CCPR/C/21/Rev.1/Add.5 www.refworld.org/docid/453883fc0.html.

UN Human Rights Committee (HRC). 2004. *CCPR General Comment No. 31: The nature of the general legal obligation imposed on States Parties to the Covenant, 26 May 2004.* UN Doc. CCPR/C/21/Rev.1/Add.13, www.refworld.org/docid/478b26ae2.html

UNDP. 2009. *Community Security and Social Cohesion Towards a UNDP Approach.* New York: Bureau for Crisis Prevention and Recovery.

UNDP. 1994. *Human development report 1994.* New York: Oxford University Press.

UNTFHS. 2016. *Human Security Handbook: An integrated approach for the realization of the Sustainable Development Goals and the priority areas of the international community and the United Nations system.*

VindingD. and MikkelsenC. 2016. *The Indigenous World 2016.* International Work Group for Indigenous Affairs.

Working Group of Indigenous Peoples in the Barents Euro-Arctic Region (WGIP). 2012. *Action Plan for Indigenous Peoples in the Barents Euro-Arctic Region 2013–2016*, Kirkenes, February 10.

Working Group of Indigenous Peoples in the Barents Euro-Arctic Region (WGIP). 2017. *Action Plan for Indigenous Peoples in the Barents Euro-Arctic Region 2016–2018*, Murmansk, March 14.

YLE. 2017. Sámi activists occupy island in protest at Tenojoki fishing rules. https://yle.fi/uutiset/osasto/news/sami_activists_occupy_island_in_protest_at_tenojoki_fishing_rules/9717663.

ZimmerbauerK. 2013. Unusual regionalism in Northern Europe: the Barents Region in the making. *Regional Studies* 47: 89–103.

2.9 Political security in the Barents Region

Dorothée Cambou

Political security is a crucial component of human security. The concept underlies and crosscuts all other aspects of human security. However, it is also poorly defined and the topic is not well addressed from a Barents perspective. In this regard, the purpose of the following analysis is to define and assess contemporary political security challenges in the Barents Region. In this assessment, the study more specifically argues that political security is intertwined with democracy and human rights but that some issues and concerns subsist in the Barents Region notably in relation to the respect and protection of the rights of citizens, the rights of indigenous peoples and the preservation of the peace and security in the region following the impacts of the Ukrainian crisis.

1 Definition

Originally the concept of political security was 'not widely used or accepted in social studies' (United Nations Development Programme 1990, 3). It is also a concept that is imprecise and ambivalent. Nevertheless, it has also been more recently considered that political society remains a 'relevant concept for gathering concerns in the political action' (Costa 2008, 568). In this regard, it is important to clarify the concept in order to ensure a more informed debate about its use and implementation.

Traditionally, political security has been loosely defined in reference to the protection of basic human rights. In this regard, political security must be distinguished from national security, which essentially focuses on the protection of the nation and the values espoused by the national society. Political security is concerned with the rights of individuals and peoples, not those of the states or nations. In its 1994 report, the UNDP indicates more precisely that political security is one of the most 'important aspects of human security' and includes the protection of people against 'political repression, systematic torture, ill treatment or disappearance' as well as 'political detention and imprisonment' (United Nations Development Programme 1990, 32). In accordance with this interpretation, political security therefore encompasses the defence of individuals against any form of political repression or human rights abuses. A predominant focus on repression is, however, insufficient in comprehensively addressing the issue of political security. While political security necessarily entails the negative obligation of the state to refrain from interfering with basic individual freedoms, it also includes positive obligations for national authorities to take necessary measures to safeguard basic political rights (Hokkanen v. Finland, 24 August 1994).[1] The protection of such rights include the freedom to vote, freedom of speech, freedom of press, the freedom of thought, conscience, religion or belief, and the freedom of movement. These rights are protected under international law, in particular the International Covenant on Civil and Political Rights (ICCPR), which has been ratified by 184 states around the world. In this sense, political security also ensures people's ability to take part in the political life of the society to which they belong therefore sustaining a democratic governance system.

In this sense, political security is closely intertwined with democratic governance: 'democracy provides the political foundation necessary to sustaining all other dimensions of security' (Young 2002). Similarly, political security involves 'support to transition to democratic practices', a view also shared by the United Nations Trust Fund for Human Security (UN Human Security Unit 2009, 21). There is also a growing consensus under international law that only in a democratic system can individuals fully exercise their human rights. Conversely, 'only when human rights are respected can democracy flourish' (European Commission 2013). Thus, human rights and democracy are inextricably related and mutually reinforcing. Democratic governance provides the framework, institutions and process to uphold a state's responsibility to safeguard the rights and needs of the population. On the other hand, human rights provide a set of values that inform the content of democratic governance and sustain political security. In practice, achieving the ambitions of political security has therefore become tied to the promotion of democracy and human rights.

The links between democratic governance and human rights are enshrined in the Universal Declaration of Human Rights and further developed in the ICCPR, which recognises a host of political rights and civil liberties underpinning meaningful democracies. More specifically, article 25 of the Covenant recognises and protects the right of every citizen 'to take part in the conduct of public affairs, the right to vote and to be elected and the right to have access to public service'. According to the Human Rights Committee, which is the body of independent experts that monitors implementation of the ICCPR by its State parties, 'Article 25 lies at the core of democratic government based on the consent of the people' (General Comment No 25, 1996, para.2). However, democracy does not only centre on the electoral process. The definition of democracy must be holistic and encompass procedural and the substantive elements. In other words, democratic governance must ensure that the outcomes of elections are representative of the people. It must also ensure that the will of peoples is freely exercised, and that all individuals can participate in political affairs free from discrimination. Democracy also implies the establishment of governance structures, which ensure freedom of expression and access to information, as well as guarantees the right to freedom of association and peaceful assembly. Democracy is thus an intersectional issue. It enjoins all political rights and freedoms necessary to ensure 'the freely expressed will of the people to determine their own political, economic, social and cultural systems and their full participation in all aspects of their lives' (UN General Assembly 2009).

Whereas democracy originally focused on the individual rights of citizens to freely participate in electoral processes without discrimination, democracy also 'entails a principle that everyone whose basic interests are affected by policies should be included in the process of making them' (Young 2002). In this regard, the concept of democracy has now become more inclusive and implies the participation of minorities and indigenous peoples groups in the decision-making processes that affect them. More particularly, ensuring the engagement of indigenous peoples and their organisations has now become critical for preventing, resolving conflicts and enhancing democratic governance (UNDP 2001). In this regard, the study of democracy as a governance framework involves the collective inclusion of indigenous peoples and minority groups in decision-making processes concerning them and their right to participate in the governance of their traditional land and natural resources. This conceptualisation of democracy is far-reaching, and is further entrenched within the intersections between political security and personal and community security (*see* chapters 2.7 and 2.8).

Alongside the extension of its subject, the extension of the concept of democracy outside the framework of the state polity is also increasingly taken into account. As postulated by Grigorescu, during these last decades, democratic norms have spread from domestic politics to intergovernmental organisations (IGOs) (Grigorescu 2015). Democratic values influence the institutional design and work of intergovernmental organisations such as the UN. In effect, democracy is now one of the universal and indivisible core values and principles of the United Nations. Thus the contemporary definition of democracy must be inclusive and comprehensive. It concerns the decision-making processes that involve formal institutions and informal processes; people of all genders, majorities and minorities; governments and civil society; and includes institutions at the national, local and international levels.

While it is 'obvious that democracy, or the installation of democratic regimes, is a component of political security' (Costa 2008, 562), democracy remains flexible and is set on open-texture political values. As noted by the UN, 'democracy is a universal value based on the freely expressed will of people'. Moreover, it is also recognised as well that, while these norms and standards are both universal and essential to democracy, there is no one model of democracy (General Assembly resolution 62/7). Indeed, the ideal of democracy is rooted in philosophies and traditions from many regions of the world. While the western standards of democracy have a value for the new democracies, the democratic ideal should integrate and suits local conditions and particularities (Beetham et al. 2008). In this regard, securing democratic governance as a means to ensure political security can be pursued through diverse routes. This diversity has cast some doubts on whether any form of democracy can always provide political security for all. It is also questionable to which extent the standards of universal democracy can/are truly being dissociated from western values in practice.

Nonetheless, because democracies are less conflict-prone than non-democracies (e.g. Russett and Oneal 2001) and democratising countries with well-managed transitions are less likely to be engaged in interstate warfare (Landman 2005), ensuring democratic governance continues to be a main element in enhancing political security. Ultimately, the promotion of democracy as a means to ensure political security must therefore be valued and contextualised in accord with the population that invokes it.

2 Contextualisation

Preserving and maintaining political security requires democratic governance. However, 'democratization is a process that requires time and patience', and which must 'be built from within societies' and 'cannot be imported, or exported, but supported' (Beetham et al. 2008, 7). As noted in the 1993 Barents declaration, the Barents institutions were established to meet this need: to support 'the ongoing process of reform in Russia which aims, *inter alia,* at strengthening democracy, market reforms, and local institutions, and which is therefore important for closer regional co-operation in the Barents Euro-Arctic Region'. After the fall of the Berlin Wall and the end of the Cold War, the situation in Russia and its relation with Nordic countries changed. During this period, the discourse on political security also shifted from traditional security, which concerns the protection of the state, military issues and war, to a focus on human security and the needs of the Barents population. With the collapse of the Soviet Union, the Russian Federation begun its transition to democracy. In line with democratising principles, the Russian Federation pursued renewed cooperation with its European neighbours, notably increasing its cooperation with Nordic states through the Barents Euro-Arctic Council. In 1996, the Russian Federation acceded to the Council of Europe, the oldest European institution and main protector of rule of law, human rights and democracy on the old continent 'on the basis of its democratic progress, taking into account the particular circumstances following the fall of the Soviet Union.' (Bindig 2010, 35; Massias 2007, 6; Stahl 2011, 176).

Since the fall of the Soviet Union, support for democratic development has become a vital aspect of Barents cooperation. According to the Norwegian Barents secretariat, the Barents cooperation is considered 'as an integral part of creating a stable, democratic and prosperous Europe', which brings 'administrative structures closer to the citizens and to improve the democratic functions of society' (The Norwegian Barents Secretariat 2017). This recognition of the common democratic values of the region are also present in the 1993 Kirkenes Declaration and in reference to several other documents published under the auspices of the Barents cooperation.[2] In 2013, the Kirkenes Declaration reiterated that the Barents cooperation is 'a unique undertaking that confirms the value of close interaction between intergovernmental cooperation, cooperation among county administrations and direct people-to-people cooperation' that seeks to develop its 'societies in full respect of internationally recognised principles for ensuring sustainable development'. Thus, the achievements of the last two decades have been the development of closer cooperation between local and regional initiatives by Barents communities and institutions, as well as the increase in activities across sectors such as business and civil society (Kirkenes Declaration 2013). In this regard, the regional framework has successfully strengthened cooperation between governance mechanisms.

Despite this progress, the democratic transition of the whole Barents Region remains incomplete. In its 2015 concluding observation report, the Human Rights Committee expressed its continuous concern regarding the Russian Government's failure to ensure freedom of expression and freedom of association, as well as to combat the rise of racism, xenophobia, Islamophobia and anti-Semitism, and to eradicate torture and ill treatment (Human Rights Committee 2015). Today, the Russian Federation also continues to have the largest number of pending cases before the European Court of Human Rights. Moreover,

> the examination of recent developments in the field democratic transition at local and regional level has shown that Russia, despite first positive signs at the beginning of the reforms in 2000, has still not been able to meet the obligations it committed to when acceding to the CoE.
>
> (Stahl 2011)

Although this issue mainly concerns Russia, it also has implications for the Barents Region. In 2017, the Norwegian Helsinki Committee, a non-governmental organisation that works to promote respect for human rights, requested 'Norwegian authorities to increase the emphasis on democracy and human rights in the Russian part of the Barents Region,' because 'on the Russian side, the human rights problems remain very severe' (Human Rights House 2007). In addition, studies like the *Freedom in the World* survey for 2017 notes a continued decline in Russia in the field of political rights and civil liberties, notably 'due to the heavily flawed 2016 legislative elections, which further excluded opposition forces from the political process' (Freedom House 2017a). Similarly, the *Nations in Transit* survey for 2017 indicates a low Russian performance when it comes to national democratic governance, electoral processes, civil society, independent media, local democratic governance, independence and corruption (Freedom House 2017b). Altogether, a low performance on these elements represent an important threat to the democracy and the political security of Russian citizens.

In contrast to the Russian Federation, the Nordic countries are among the champions of human rights and democracy. For many years, Norway, Sweden and Finland have shared the top positions on several democracy analyses (Global Democracy Ranking 2015; The Economist 2016; Freedom House 2017a). Yet, the Nordic countries are not entirely exempt from political insecurity concerns. The consolidation of democracy and the safeguard of human rights are continuous processes. As indicated by minister of foreign affairs Børge Brende, human rights and rule of law are central components of good democratic governance:

> human rights provide a framework for identifying and addressing inequalities and thus ensuring that no one is left behind. This includes safeguarding the rights of those who are hardest to reach – the most vulnerable and marginalised groups, including indigenous peoples and persons with disabilities.
>
> (Brende 2014)

The recognition and protection of the rights of the indigenous peoples in the Nordic counties remains nonetheless a major issue. The protection of the rights of the Sámi people, and their inclusion in the democracy of their respective states, implies the recognition and implementation of their right to self-determination. This has yet to occur in practice (Anaya 2011; Tauli-Corpuz 2016). Whereas the economic and social situation of the Sámi people is better than it is for many other indigenous peoples around the world, the Sámi in the Nordic countries continue to face major human rights challenges that affect their political security as a distinct people living in western democratic states (Multiculturalism Policy Index 2010).

Finally, political security is interconnected with peace and stability. However, the annexation of Crimea in 2014 and the war in Eastern Ukraine have raised questions about the Russian Federation's governmental relationship with European countries. This event also incurred great speculations about conflict in the Barents Region that could threaten the political security of its population.

With this context in mind, the purpose of the next section is to assess some of the issues that currently challenge the democratic order of the Barents Region and the political security of its population.

3 Assessment

There are many political security challenges that can be identified in the Barents Region. While such challenges are complex and interconnected, it is possible to narrow the scope of insecurity issues to three main components. The first component concerns the protection of democratic rights, which includes safeguards on the rights of individual citizens to vote, and their other political freedoms. The second component concerns the collective rights of indigenous peoples, their inclusion in decision-making process and the exercise of their right to self-determination. The last component concerns the stability of the region and the geopolitical relations between the Russian Federation and its neighbouring countries, as well as the preservation of cooperation across the borders of the Barents Region.

The following section subsequently analyses each component as a challenge to political security in the Barents Region.

3.1 Political rights and freedoms

In the Barents Region, the consolidation and strengthening of democracy is an on-going process. This process entails the protection of the right to vote and the guarantee of political freedoms at all levels of the decision-making process.

The right to vote in national, regional and local elections is crucial to ensure the establishment of representative institutions in a democracy, because the authority of the government derives itself from the consent of the governed. According to statistics from Patchwork Barents,[3] the voter turnout in the latest elections to national, regional and local legislative assemblies, as well as in regional parliament elections, varies substantially within the Barents Region. National elections are much more popular among Barents citizens, with a turn out ranging from 50% in Argnaglesk Oblast to 87.5% in Vasterbotten (Ulyanova 2015). Also, whereas Sweden had a voter turnout of over 80% in regional elections, all other regions in Norway, Finland and Russia registered considerably lower participation. In Arkhangelsk Oblast, only 20.99% of the electorate participated in the gubernatorial elections and in Murmansk Oblast, the average turnout throughout the regional elections was 23.10%. In contrast, elections held in the Republic of Karelia in 2011 gathered a turnout of 44.4%, and in the Komi Republic in 2015, this figure was 44.2%. At the regional level, Finnish and Norwegian voters were also more active voters. The latest (2012) turnout figures for Lapland and Northern Ostrobothnia were 60.6 and 56.5%. In Norway, according to NRK,[4] the local election turnout in Nordland, Troms and Finnmark was 59.8, 58.9 and 58.1% respectively in 2015. In comparison with previous elections, the turnout in Norway was, however, significantly lower than in the previous elections organised in 2011, which gathered a turnout of 63.4, 64.6 and 61.7% voters across the three counties respectively. Thus, the participation of citizens in the electoral process has been decreasing, though this is probably also reflective of a global crisis of trust in the democratic process.

However, while all modern democracies hold elections, not all elections are democratic. In order to ensure democratic elections, all citizens should be entitled to exercise their right to vote free from discrimination. Genuine and periodic elections are also essential to ensure the accountability of representatives for the exercise of the legislative or executive powers vested in them. In accordance with Article 25 ICCPR, ‘such elections must be held at intervals which are not unduly long and which ensure that the authority of government continues to be based on the free expression of the will of electors.' According to the HRC, reasonable limitations on campaign expenditure may also 'be justified where this is necessary to ensure that the free choice of voters is not undermined or the democratic process distorted by the disproportionate expenditure on behalf of any candidate or party' (HRC 1996, para. 19).

In the Russian Barents, however, irregularities have been noted which question the democratic nature of national and local elections. According to the independent election watchdog, Golos,[5] 'the main problems in Russian elections are connected with the activities of state officials on both national and local level. The officials actively use their available administrative resources to promote party interests in public events and the media' (Staalesen 2016). These irregularities were especially reported in Arkhangelsk, which was one of the regions with the biggest number of registered offences.

Another issue of concern in the Barents Region concerns youth. Whereas the interests of young people and their involvement in different regional activities is essential to strengthening civil society and democracy in the Barents Region,[6] the lack of engagement of youth in political affairs is sufficiently significant to be reported. In Norway, it has been noted that the number of first-time voters (18–21 years) practising their right to vote in local elections dropped from 53% in 1971 to below 30% in 2007 (Dalhaug 2012, 77). Voter turnout among youth is also lower than among adults in both local and national parliament elections, and the number of young members in Norwegian political parties has also fallen from about 44.000 in 1977 to 23.000 in 1995.[7] Russia is experiencing similar disproportionality, with figures in 2007 from the Komi Republic showing that 37.3% of the population aged 18–30 years took part in the regional elections, 49.6% in elections to the State Duma and 58% in presidential elections. While the lack of political engagement in young people through participation in elections should not be overstated, according to Dalhaug, project manager for the Barents Youth project at the Norwegian Barents in 2003, 'the low number of young people involved in political activity in more conventional forms calls for concern' (Dalhaug 2012, 77). More particularly, the low number of political youth organisations and the low number of youth representatives in local and regional constituencies can undermine the representative outcome of political decision-making processes and their relevance for youth. As noted by the Barents Youth Council, 'youth need experience in democratic decision-making processes to understand the benefits of democratic societies' (BEAC Youth Working Group 2011). In this regard, youth involvement in the Barents Region must be strengthened through their inclusion in decision-making processes and through increased support for youth engagement in NGOs and media, in order to foster active citizenship and a plural democratic process.

Finally, participation in electoral processes is no panacea for democracy. Beyond participation in the political processes, it is also important to ensure that all citizens benefit from their political freedoms. However, the democracy scores and regime ratings of the country of Russia continue to raise general concern, which also affects the political security of the Barents population. According to the Freedom of House's 2017 report, Russia's democratic score is declining. More particularly, the organisation argues that 'the regime sought to perfect its authoritarian control over Russian citizens and the economy in order to prevent any real democratic changes' (Freedom House 2017b). To some extent, those conclusions are consistent with the 2014 HRC conclusions on Russia's State report, which stresses the need for the government to take all necessary measures to ensure that individuals fully enjoy their freedom of speech and the expression of dissenting political opinion, their right to peaceful assembly and their freedom of association (HRC 2015, para 18–22). In particular, government measures to suppress political dissent and the 'reports of harassment, violence and killing of lawyers, journalists, human rights defenders and opposition politicians' have raised strong concern in Russia (HRC 2015, para 18–19). In relation to peaceful assembly, the HRC also expressed concern

> about consistent reports of arbitrary restrictions on the exercise of freedom of peaceful assembly, including violent and unjustified dispersal of protesters by law enforcement officers, arbitrary detentions and imposition of harsh fines and prison sentences for the expression of political views.
>
> (HRC 2015, para. 21)

Concerning the freedom of association, the Committee also worried about 'restrictions on the operations of NGO activities and to suspension or voluntary closure of some NGOs' (HRC 2015, para. 22). The right to free association and peaceful assembly are, however, essential in conjunction to the rights of individuals to participate in political affairs, and are protected under article 25 of the ICCPR. In this regard, it is required of states to ensure that basic freedoms are protected in their internal management, in order to enable citizens to fully exercise their democratic rights. In the absence of fully-fledged protection of such rights, the political security of citizens living in the Russian Federation remains consequently pale in comparison to the situation afforded to the population in the Nordic countries.

3.2 Inclusion and self-determination

While it is fundamental to ensure that individual citizens can fully exercise their democratic rights, it is also crucial to ensure that all peoples participate in political life and are active decision-makers – especially indigenous peoples. This is essential to overcome historical inequalities and discrimination. However, in the Barents, the inclusion of indigenous peoples in the democratic system remains an ongoing issue in all four states.

In the Nordic states, the recognition and protection of the rights of the Sámi people has made some progress in the last thirty years. This includes the establishment of a Sámi assembly in each of the three Nordic countries to represent the Sámi community at the domestic level, which also exercises certain competencies in the decision-making process affecting the Sami livelihoods. It also includes the adoption of multiple national regulations and legislations to ensure the protection of the cultural rights of the Sámi people, their language and traditional way of life. Yet, such progress has not been sufficient to remediate the historical inequality the Sámi people have faced, and have not evolved to protect them from new challenges, such as those provoked by the negative impacts of industrial development on their traditional lands. While it is important to reform legislative and political frameworks to guarantee the human rights of the Sámi people, the autonomy and self-governance powers of the Sámi parliaments must also be strengthened in order to ensure the right of the Sámi people to self-determination.

The right of the Sámi people to self-determination is an essential feature of their human rights and is the basis for their inclusion in the democratic orders that govern them. As noted by both the former and current Special Rapporteur on the Rights of Indigenous Peoples, there is an ongoing need to strengthen the Sámi Parliament's ability to participate in and genuinely influence decision-making in matters that affect the Sámi people (Anaya 2011, para. 37). In Finland and Sweden for instance, the Sámi Parliament does not have substantial influence or decision-making powers. Whereas in Finland the Sámi parliament has limited advising powers, in Sweden, the Sámi Parliament functions as both a popularly elected body and a State administrative agency which includes an obligation of the Sámi parliament to implement policies and decisions made by the Swedish Parliament and government institutions. This can be problematic, however, when decisions made by the Swedish Parliament do not reflect the view and interests of the Sámi people. In Norway, substantial progress has been made, especially since the Norwegian government has ratified Convention ILO 169 concerning Indigenous and Tribal Peoples, and adopted a consultative agreement in 2005 to strengthen the political influence of the Sámi Parliament in political processes (Falch et al. 2016). Yet, in its 2016 report, the current Special Rapporteur, Tauli-Corpuz, noted that Norway should 'enhance efforts to implement the right of the Sámi people to self-determination and to more genuinely influence decision-making in areas of concern to them' (Tauli-Corpuz 2016, para. 76). In particular, she advised the need for a more effective consultation arrangement, which should also cover budgetary decisions. The following recommendations have, however, not yet been addressed by the Norwegian government.

In Russia, indigenous peoples have also called for 'greater democracy and participation' (Prakhova 2005). In effect, the situation for the Sámi people and other indigenous communities living in the country is even more alarming than it is in the Nordic states. This is in spite of the guarantees provided by the Russian Federation for the rights of small indigenous peoples under Article 69 of the Constitution. In its 2014 report, the HRC indicated for instance that it remains concerned regarding insufficient measures taken to respect and protect the rights of indigenous peoples, and to ensure that members of such peoples are recognised as indigenous in Russia. In its report, the HRC stressed particular emphasis on the lack of protections for the traditional lands and sacred areas of indigenous peoples, which are 'unprotected from desecration, contamination and destruction by extractive, development and related activities'. In addition, it also emphasised 'that consultation with indigenous peoples on matters of interest to their communities is insufficiently enforced in practice and that access to effective remedies remains a challenge' (HRC 2015, para. 23). This is the case for Nenets in Russia, where the development of extractive industry operations continues to undermine the traditional way of life of Nenets people, and activities continue without obtaining the free, prior and informed consent of the concerned communities for most proposed projects. Thus, there is a need for new governmental measures in order to effectively ensure the accommodation of the rights of indigenous peoples at the domestic level, and especially to ensure their right to self-determination, including their right to fully and meaningfully participate in the management of their land and territories.

Additionally, this extends to the right of indigenous peoples to represent their own interests through their own and truly representative organisations. The closure and reform of RAIPON, the Russian Association of Indigenous Peoples of the North, Siberia and Far East, because of an 'alleged lack of correspondence between the association's statutes and federal law' have sparked major reactions both regionally and internationally. Today, it is still questioned whether the organisation operates under free and genuine indigenous control (International Work Group for Indigenous Affairs 2012). In the absence of representative institutions, the situation of indigenous peoples' rights is at risk of further erosion in Russia.

Finally, while the protection of the rights of indigenous peoples occurs primarily at the domestic level, it is also important for indigenous representatives to be included in the decision-making processes that operate on regional and international levels. In the Barents Region, indigenous peoples have strengthened their influence in decision-making processes affecting them through their participation in the Working Group on Indigenous Peoples. The working group consists of the representatives of the Sámi, Nenets and Veps peoples. It was established as a means to recognise the specific needs and interests of indigenous peoples living in the Barents Region. The working group has a specific position within the Barents cooperation framework; in addition to its operational role as a working group, it also has an advisory role to both the Barents Euro-Arctic Council (BEAC) and the Barents Regional Council (BRC) (Working Group of Indigenous Peoples – BEAC 2017). Through this advisory function, the working group consequently has a political dimension. It can influence the decisions made by the BRC and the BEAC to the extent that the councils take its opinions into account. In addition, the working group can also participate in all Barents working group sessions. However, the working group solely offers consultative status for indigenous peoples within the Barents Councils institutions. Therefore, in order to enhance their rights to self-determination, the Sámi together with the Nenets and the Veps have asked for permanent status within the Barents cooperation framework, as opposed to simply being granted consultative status as a working group. The status of permanent participant would afford them direct representation instead of indirect representation through a working group (BEAC Newsletter 2010)· This way, indigenous peoples would be granted full and active participation within the BEAC and BRC as permanent participants, and, as they argue, this status would be in line with their right to self-determination.

The above analysis makes explicit that the inclusion of indigenous peoples in democratic governance remains an on-going issue in the Barents Region, both at the domestic and regional level. In order to ensure political security for all peoples in the Barents Region, it is consequently important that governance institutions at the local, national and regional levels recognise their land and cultural rights, promote inclusion and foster their full self-determination and participation in democracy as valued members of national societies and the regional Barents community.

3.3 Peace and stability

Political security is conditional to peace and stability. One of the best indicators for political insecurity in a country is the priority the government accords to military strength (UNDP 1994, 32). During the Cold War, the Barents Region was an area of military confrontation and therefore possessed a high level of human insecurity. With the collapse of the Soviet Union, the possibilities for military conflict to take place in the region declined. The development of cooperation in the region also largely contributed to enhancing the peace and stability of the region. However, although it is unlikely that a conflict could start in the region, there are concerns that the region could be affected by geopolitical tensions elsewhere. In particular, Russia's annexation of Crimea, and its involvement in the conflict in Ukraine in 2014, have resulted in a strained relationship between the Nordic countries and the Russian government, which could threaten the peaceful state of affairs in the Arctic (Rahbek-Clemmensen 2017). Although it is difficult to assess the full impact of such tensions for the Barents Region, the following subsection attempts to address some aspects relating to the peace and stability of the region.

As mentioned elsewhere in this book, the Ukrainian conflict increased political tensions between all Arctic countries. After the annexation of Crimea, the United States and the European Union enacted sanctions on Russian individuals and businesses, which were also reciprocated by Russian counter measures. Several types of sanctions have been imposed on Russia. The sanctions include asset freezes and travel bans on individuals and entities that have been involved or have benefited from Russia's actions in Ukraine, sectoral sanctions targeting the oil and gas, defence and financial sectors in Russia, and an arms embargo restrictions on doing business and investing in Russian-occupied Crimea. In addition, the sanctions also prohibit American and EU companies from providing equipment and expertise to complex deep-water, Arctic and shale oil development projects in Russia. While the impact of the sanctions on Russian resources business remains limited – gas production and exports have fallen slightly and oil production remains unaffected for the time being[8] – the impact of financial sanctions has been immediate and significant. In particular, sanctions contributed to the collapse of the Russian ruble and the Russian financial crisis. With the impact of the 2008 financial crisis, such sanctions have isolated Russia from the global economy and held back its economic modernisation (EU Parliament 2016). Furthermore, it has also been reported that sanctions caused economic damage to a number of EU countries, with total losses estimated at €100 billion. In the Barents Region, Norway was particularly affected by the sanctions on Russia, with a drop by 28% year-on-year to a total of €672 million in the first half of 2015 and a major export decline, though Russian trade to Norway in the period increased by 10% to a total of €586 million (Staalesen 2015).

Despite economic tensions and little political contact since 2014, cooperation between Norway and Russia in the north has continued at the regional level. Most notably, the Barents cooperation has continued to operate in areas such as fisheries, border control, environment and nuclear safety. Broader people-to-people contact networks between Murmansk, Arkhangelsk and the northern parts of the Nordic countries have also been maintained. In fact, maintaining good relations in the Barents Region despite tensions in bilateral relations has been an objective for all governments. The importance of preserving cooperation in the Barents Region despite major geopolitical turmoil was also noted by the head of the Barents Regional Council, Arkhangelsk Governor Igor Orlov, who 'told his Oblast government in September 2014 that complicated geopolitics should not affect Barents Cooperation' (Nilsen 2017). In practice, the organisation of a series of high-level talks in the Barents-Euro Council after the annexation of Crimea underscores that mutual interests in the region have continued to guide cooperation in the region. As such, while media headlines speculated great potential for conflicts in the Barents-Arctic, a common understanding of joint interests in maintaining dialogue has contributed to preserving the peace and stability within the region.

In addition, military activities within the region do not currently constitute a threat for its stability. Even though the activities of NATO and Russia outside the Barents have raised some concerns, it must also be noted that NATO does not have a strong presence in the Arctic. After the end of the Cold War, NATO's focus gradually shifted away from the Arctic to strengthen its role in the Mediterranean and more recently in the Eastern part of Europe. While a significant part of the region is within the territory of NATO members, NATO does not have a specific Arctic policy and therefore no clear role in the region. Nonetheless, following the Ukrainian crisis, the possibility and the importance of enhancing the presence of NATO in the region has been questioned. According to Øystein Bø (2016), State Secretary of the Norwegian Ministry of Defence, a strong NATO presence in the North would be in the

US' and Norway's interests. On the other hand, it has been stressed that the purpose of strengthening NATO's presence in the Barents does not inevitably aspire to revive or hearten conflict in the region. As noted by NATO Secretary General Jens Stoltenberg in his 2016 speech, 'NATO does not seek confrontation with Russia'. NATO 'does not want a new Cold War' nor 'wants a new arms race'. Rather, its 'aim is a more positive and a more cooperative relationship with Russia' (Stoltenberg 2016). In this regard, it is also suggested that an increasing NATO presence does not necessarily mean an increased focus on hard security issues. Instead, it has been suggested that NATO would focus on 'soft security' issues including the ecological consequences of global warming and the risk of manmade and environmental disasters (Sergunin 2015, 110). Although this focus does not exclude military components of NATO policy, as reflected in a series of exercises conducted under the NATO allegiance, Sergunin argues that NATO reinforced involvement in the Arctic would principally target human activities and the global competition for resources (Sergunin 2015, 110). As such, the strengthening of NATO presence and military activities would not necessarily constitute a factor of insecurity for its population but perhaps could provide a means to increase political security in the region.

Yet, relations between Russia and the Nordic countries remain complicated and influenced by larger geopolitical developments, especially in the domain of military activities occurring outside the Barents Region. In 2013, Russia's Prime Minister, Dmitry Medvedev, indicated during the Barents Summit in Kirkenes 'that Moscow sees the Arctic as an area with good opportunities to implement joint programs and initiatives' (Nilsen 2017). However, he also underlined that 'any expansion of NATO to include Sweden and Finland would upset the balance of power and force Russia to respond'. With Finland and Sweden strengthening their ties to NATO and growing speculation around joining the Alliance, Russia continues to warn that they would be forced to take a military response if such a situation materialised, without however, clarifying the extent to which such a response would entail (O'Dwyer 2016). Currently, it also remains questionable whether Sweden and Finland would compromise their non-alignment policy. In 2016, Margot Wallstrom, Foreign Minister, indicated that these policies have served them well until today and 'contributes to stability and security in northern Europe' (Wallstrom 2016). However, Sweden military exercise with US and NATO troops around the Baltic Sea island of Gotland in 2017 in response to large scale military exercises by Russia on the borders of Europe have again raised the possibility of the country joining NATO. According to the Swedish Defence Minister, the Swedish military exercise constitutes 'important signal to the Swedish population and also to other countries and partners that (they) take this security situation seriously' (Hultqvist 2017). Although those events demonstrate that the security situation in the vicinity of the Nordic states have deteriorated since the Crimea crisis (Fiskvik 2016), it nevertheless remains difficult to prognosticate to which extent those developments will affect more generally the stability of the Barents Region and its population.

In any event, ensuring peace and stability requires the maintenance of dialogue between all national governments, in particular concerning military activities in order to ensure that regional military activities do not jeopardise cooperation in the Barents. In addition, it is also important to ensure transparency in military exercises along national borders and to ensure the provision of necessary information to all concerned parties. An increase in Russian unannounced military exercises since 2014 have raised concerns regarding the Russian government's lack of clear expression on its motivations (NATO's Timo Koster and US deputy permanent representative to NATO Earle Litzenberger in Gronning and Norwegian Institute of Foreign Affairs Conference 2016). In this regard, it is generally agreed that more 'stability, predictability and cooperation with Russia is needed in the Arctic' (Gronning and Norwegian Institute of Foreign Affairs Conference 2016). Whether these goals are achievable with a more involved NATO in the region is, however, still contested (Gronning and Norwegian Institute of Foreign Affairs Conference 2016). As a result, monitoring the development of peaceful military relations and activities in the Barents should continue in order to ensure that cross-border cooperation and the promotion of human security prevail in the future.

4 Conclusions

With the establishment of the Barents cooperation and the development of cross-borders relations between peoples and institutions since the 1990s, the Barents Region has become one of the most peaceful and stable regions in the world.

Yet, the political security of the Barents population is not unchallenged. As previously discussed, beyond peace and stability, political security necessarily involves guaranteeing the democratic governance of states and human rights of the population. In the Barents Region, there is an excellent level of protection afforded to the individual political rights of citizens living in the Nordic countries, but mediocre guarantees provided for those living in Russia. There are also concerns in both countries about the level of participation in electoral activities, which constitutes the primary medium to ensure the exercise of democracy. Finally, there is also a clear lack of political freedom in Russia and a lag in the rights to free press, association and assembly, which in turn constrain democracy and political security in this part of the region. In this regard, there is a clear and urgent need for the promotion, enhancement and assurance of basic human rights and political freedoms in the Barents Region for all citizens.

Additionally, the present analysis has also demonstrated that political security in the Barents Region concerns the inclusion of indigenous peoples in the political and democratic order of the Barents Region. For many decades, indigenous peoples have been colonised, marginalised and forcibly assimilated by their states. Today, they continue to suffer from the grievous consequences of historical injustices and contemporary challenges, which include the lack of inclusion in decision-making processes affecting them and the exploitation of their traditional lands and resources. Fuelling more support to indigenous representative institutions through the recognition of their rights to self-determination is therefore fundamental to increasing political security in the Barents Region. In particular, the ratification of ILO Convention 169 and the adoption of the Nordic Sámi Convention would be key elements to enhance the political and legislative framework of all the Barents states concerning the rights of the Sámi people. While the Barents institutions have already recognised the specific status of indigenous peoples through the establishment of the working group on indigenous peoples, strengthening the role and influence of this mechanism would also be a means to increase the inclusion of indigenous peoples in the decision-making processes affecting them at the regional level. Ultimately, there is a pressing need to adopt a holistic perspective towards the rights of indigenous peoples in order to ensure that their political security is equally ensured in the Barents Region.

Finally, preserving peace and stability in the region is also required. Political security remains indeed conditional to peace and stability. Whereas it is unlikely that conflicts would resurface in the Barents Region (Rahbek-Clemmensen 2017), reinforcing dialogue and cooperation between concerned actors and strengthening transparency and predictability in the field of military activities remains necessary both within and outside the Barents Region. While cooperation in the Barents Region has been maintained despite the Ukrainian crisis, recent events have raised some tensions and concerns regarding military relations and its significance for the region. In this context, maintaining peaceful military relations between states must continue in order to ensure that cross-border cooperation and the improvement of human security prevail in the region.

Notes

1 European Court of Human Rights, Hokkanen v. Finland, 24 August 1994.
2 XII Session of the Barents Euro-Arctic Council, Joint communique,Murmansk, 15 October 2009; XV Session of the Barents Euro-Arctic Council, Oulu, 15 October 2015.
3 Patchwork Barents, http://patchworkbarents.org/, last visited on 17-07-2017
4 NRK, Norwegian Result, www.nrk.no/norge/resultater-valg-2015-1.12499895, last visited on 17-07-2017.
5 Golos was awarded the Sakharov Prize its 'outstanding efforts to promote democratic values through free and fair elections' by the Norwegian Helsinki Committee in 2012.
6 BEAC Committee of Senior Officials 2008.
7 Norwegian Government, NOU 2001: 03, www.regjeringen.no/nb/dep/krd/dok/nouer/2001/nou-2001–03.html?id=143453, last visited on 17–07–2017.
8 However, it is stated that oil export earnings have dropped significantly but this is certainly due to lower oil prices. In 2016, Russian oil companies were planning to export 6% less than in 2015, (EU Parliament, 2016).

References

Anaya, S.J. 2011. *Report of the Special Rapporteur on the Rights of Indigenous Peoples: The Situation of the Sami People in the Sápmi Region of Norway, Sweden and Finland.* UN Doc. A/HRC/18/35/Add.2. New York: Human Rights Council.

BEAC. 2013. Kirkenes Declaration 2013.

BEAC Newsletter. 2010. Increased status for indigenous peoples in the Barents cooperation? *Barents Euro-Arctic Cooperation.* www.barentscooperation.org/newsletter/Newsletter-1-2010/Increased-status-for-indigenous-peoples-in-the-Barents-cooperation.

BEAC Youth Working Group. 2011. *The Barents Regional Youth Program 2011–2014, Appendix 1.* www.barentsinfo.fi/beac/docs/The_Barents_Regional_Youth_Program_2011_2014_Draft_Appendix_1.pdf.

Beetham, D.*et al.*2008. *Assessing the Quality of Democracy. A Practical Guide.* Stockholm: International Institute for Democracy and Electoral Assistance.

Bindig, R. 2010. Russia's accession to the Council of Europe and the fulfilment of its obligations and commitments. In *Russia and the Council of Europe: 10 Years After,* edited by K. Malfliet and S. Parmentier. Houndmills: Palgrave Macmillan.

Bø, Ø. 2016. Opening remarks at NADIC by Øystein Bø. www.regjeringen.no/en/aktuelt/opening-remarks-at-nadic-by-oystein-bo/id2549269/.

Brende, Børge 2014. Nordic statement on human rights and good governance. Taleartikkel. *Regjeringen.no.* June 9. www.regjeringen.no/no/aktuelt/nordisk_innlegg/id763330/.

Costa, G. 2008. Political security, an uncertain concept with expanding concerns. In *Globalization and Environmental Challenges,* edited by H.G. Brauch and Ú. Oswald Spring, 561–568. Hexagon Series on Human and Environmental Security and Peace 3.Berlin: Springer. http://link.springer.com/chapter/10.1007/978-3-540-75977-5_42.

Dalhaug, L. 2012. Children and youth as a priority, Barents review 2012. In *Barents Borders Delimitation and Internationalization,* edited by A. Staalesen. Barents Review. The Norwegian Barents Secretariat.

EU Parliament. 2016. Sanction over Ukraine impact on Russia. Briefing. EU Parliament. www.europarl.europa.eu/RegData/etudes/BRIE/2016/579084/EPRS_BRI(2016)579084_EN.pdf.

European Commission. 2013. Democracy and human rights policy – democracy – international cooperation and development. *International Cooperation and Development.* https://ec.europa.eu/europeaid/sectors/human-rights-and-governance/democracy-and-human-rights/democracy_en.

Falch, T., Selle, P. and StrømsnesK. 2016. The Sámi: 25 years of indigenous authority in Norway. *Ethnopolitics* 15(1): 125–143.

Fiskvik, J. 2016. Nordic security: moving towards NATO? In *CSS Analyses in Security Policy No. 189,* edited by C. Nünlist. Zurich: Center for Security Studies (CSS).

Freedom House. 2017a. Freedom in the world 2017. https://freedomhouse.org/report/freedom-world/2017/russia.

Freedom House. 2017b. Nations in transit 2017. https://freedomhouse.org/report/nations-transit/2017/russia.

Global Democracy Ranking. 2015. Democracy Ranking 2015. *Democracy Ranking.* http://democracyranking.org/wordpress/rank/democracy-ranking-2015/.

Grigorescu, A. 2015. *Democratic Intergovernmental Organizations?: Normative Pressures and Decision-Making Rules.* Cambridge: Cambridge University Press.

Gronning, R. and Norwegian Institute of Foreign Affairs Conference. 2016. High North News – NATO reluctant to engage in the Arctic. *High North News*. www.highnorthnews.com/nato-reluctant-to-engage-in-the-arctic/.

Hultqvist, P. 2017. Sweden gears up for biggest military exercise in decades. *Financial Times*. www.ft.com/content/11e9a55c-93b3-11e7-a9e6-11d2f0ebb7f0

Human Rights Committee. 1996. General Comment No. 25: The right to participate in public affairs, voting rights and the right of equal access to public service (Art. 25). UN Doc. CCPR/C/21/Rev.1/Add.7.

Human Rights Committee. 2015. *Concluding Observations on the Seventh Periodic Report of the Russian Federation*. UN Doc. CCPR/C/RUS/CO/7.

Human Rights House. 2007. Politics towards the north must prioritise democracy and human rights. http://humanrightshouse.org/Articles/5773.html.

International Work Group for Indigenous Affairs. 2012. Why the Russian Government Shuts down RAIPON – Background Article. *IWGIA*. www.iwgia.org/news/search-news?news_id=710.

Landman, T. 2005. Democracy and human security: essential linkages. In *Democracy, Conflict and Human Security: Pursuing Peace in the 21st Century*, edited by J. Large and T. D. Sisk. Stockholm: International IDEA.

Massias, J.P. 2007. Russia and the Council of Europe: ten years wasted? *Russie. Nei. Vissions* 15. www.ifri.org/en/publications/enotes/russieneivisions/russia-and-council-europe-ten-years-wasted.

Multiculturalism Policy Index. 2010. *Multiculturalism Policy in Contemporary Democracies*. Kingston, Canada: Queen's University. www.queensu.ca/mcp/.

Nilsen, T. 2017. Barents cooperation in winds of change. http://arcticyearbook.com/commentaries2014/127-barents-cooperation-in-winds-of-change.

O'Dwyer, G. 2016. Russia issues fresh threats against unaligned Nordic states. *Defense News*. www.defensenews.com/story/defense/international/2016/05/05/russia-issues-fresh-threats-against-unaligned-nordic-states/83959852/.

Prakhova, A. 2005. Call for greater democracy and participation. www.norden.org/en/news-and-events/news/call-for-greater-democracy-and-participa.

Rahbek-Clemmensen, J. 2017. The Ukraine crisis moves north. Is Arctic conflict spill-over driven by material interests? *Polar Record* 53(1): 1–15.

Russett, B.M. and Oneal, J.R. 2001. *Triangulating Peace: Democracy, Interdependence, and International Organizations*. New York: Norton

Sergunin, A. 2015. *Russia in the Arctic: Hard or Soft Power?*New York: Columbia University Press.

Staalesen, A. 2015. Trade with Russia plummets. *The Independent Barents Observer*. http://barentsobserver.com/en/business/2015/08/trade-russia-plummets-06-08.

Staalesen, A. 2016. Election violations on the map. *The Independent Barents Observer*. https://thebarentsobserver.com/en/life-and-public/2016/09/5-years-mapping-election-violations.

Stahl, A.K. 2011. Russia's path to federalism and democracy under the influence of the Council of Europe, abstract. *L'Europe En Formation* 359: 59–81.

Stoltenberg, J. 2016. The Warsaw Summit: strengthening NATO in turbulent times. www.nato.int/cps/en/natohq/opinions_131724.htm.

Tauli-Corpuz, V. 2016. *Report of the Special Rapporteur on the Situation of Human Rights and Fundamental Freedoms of Indigenous People*. UN Doc.A/HRC/33/42/Add.3. New York: Human Rights Council.

The Economist. 2016. EIU Democracy Index 2016. *Infograpics Economist*. https://infographics.economist.com/2017/DemocracyIndex/.

The Norwegian Barents Secretariat. 2017. The Barents Region. https://barents.no/en/barents-region-0.

Ulyanova, A. 2015. Swedes are most active voters in Barents Region. *The Independent Barents Observer*, September 30. http://barentsobserver.com/en/politics/2015/09/swedes-are-most-active-voters-barents-region-30-09.

UN General Assembly. 2009. *Strengthening the role of the United Nations in enhancing periodic and genuine elections and the promotion of democratization*. UN Doc. A/RES/64/155.

UN Human Security Unit. 2009. *Human Security in Theory and Practice An Overview of the Human Security Concept and the United Nations Trust Fund for Human Security*. New York. www.un.org/humansecurity/sites/www.un.org.humansecurity/files/human_security_in_theory_and_practice_english.pdf.

UNDP. 1990. *Human Development Report 1990*. New York: Oxford University Press.

UNDP. 1994. *Human Development Report 1994*. New York: Oxford University Press.

UNDP. 2001. *UNDP and Indigenous Peoples: A Policy of Engagement*. www.undp.org/content/undp/en/home/librarypage/environment-energy/local_development/undp-and-indigenous-peoples-a-policy-of-engagement.html.

Wallstrom, M. 2016. More Swedes are for joining NATO than against. *Radio Sweden*. http://sverigesradio.se/sida/artikel.aspx?programid=2054&artikel=6426670. Working Group of Indigenous Peoples – BEAC. 2017. *Barents Euro-Arctic Cooperation*. www.barentscooperation.org/en/Working-Groups/Working-Group-of-Indigenous-Peoples.

Young, I.M. 2002. *Inclusion and Democracy*. Oxford: Oxford University Press.

2.10 Digital security in the Barents Region

Mirva Salminen

This chapter depicts the current state of digital security in the Barents Region. Basing on a human security approach, it analyses the situation under the following six themes: (1) access to digital networks, (2) availability of digital services, (3) access to (relevant) information, (4) digital awareness and skills, (5) digital inclusion policies and protection of human rights, and (6) state of cybercrime and digital abuse. Digital security in the region varies and additional research is required for making well-based decisions that better include the interests, needs and wants of people in the future digital development.

1 Definition

No univocal definition of 'digital security' exists. In addition, the term is often used interchangeably with such concepts as 'information security' or 'cybersecurity'. Even if the latter terms can also be defined in multiple ways, they are both used by the United Nations[1] and hence utilised in this chapter for defining digital security as an aspect of human security in the Barents Region.

Digital security is not a dimension of human security as specified in the original UNDP[2] Human Development Report in 1994. However, over the past decades virtually all aspects of human life have been digitised to the extent that the everyday security of individuals and communities arguably depends on the uninterrupted functioning of information and communication technologies (ICTs). The realisation that information infrastructure is critical for a smoothly functioning society has led to the adoption of information and/or cybersecurity agendas throughout the globe, including the Barents Region.[3]

Digital security found its way onto the agenda of the UN General Assembly from a Russian initiative in 1998 under the heading of 'developments in the field of information and telecommunications in the context of international security' (UNODA n.d.; *see also* UNGA 1999; cf. IISS 2015, 26). Since then, the UN has had three streams for discussing digital security, namely: (1) the politico-military stream focusing on the questions of cyber peace and warfare, (2) the economic stream focusing on the criminal misuse of ICTs, and (3) the broader internet governance stream. Yet, the organisation's activities have been fragmented due to member states' competing agendas (Maurer 2011,

15–16, 18; *see also* UNODA 2015). The International Telecommunication Union (ITU)[4] with its assigned responsibility for ICT issues has often stood at the heart of these disputes (e.g. Kwalwasser 2009, 509–514).

ITU defines cybersecurity on its website as:

> the collection of tools, policies, security concepts, security safeguards, guidelines, risk management approaches, actions, training, best practices, assurance and technologies that can be used to protect the cyber environment and organization and user's assets.[5] [...] The general security objectives [in the cyber environment] comprise the following: availability, integrity (which may include authenticity and non-repudiation), and confidentiality.[6]

Confidentiality, integrity and availability of information are building blocks of the majority of information security conceptualisations. For example, SANS Institute's[7] definition as 'the processes and methodologies which are designed and implemented to protect print, electronic, or any other form of confidential, private and sensitive information or data from unauthorized access, use, misuse, disclosure, destruction, modification, or disruption,'[8] repeats similar conceptual elements.

Whereas cybersecurity gravitates towards the maintenance of national security in the digital environment and international cooperation in countering cybercrime, the conceptualisation of digital security in relation to the daily lives of individuals and communities receives less attention. Therefore, by naming the object of study as 'digital security'[9] and highlighting its importance for individual and community wellbeing in the Barents Region, this chapter aims at integrating digitalisation firmly with human security. It recognises individuals and communities as actors who actively impact (in)security and (un)trustworthiness of the digital environment and, thus, the everyday life of themselves and others.

2 Contextualisation

A trustworthy and secure digital environment is necessary for a smoothly functioning society in the Barents Region. It enables the utilisation of social, economic, cultural, educational and informational opportunities provided by digitalisation to all stakeholders. However, in comparison to the main urban areas of Norway, Sweden, Finland and Russia, the information infrastructure in the Barents Region is thin (sparse fibre optic cable network and highlighted reliance on mobile technologies), unstable, perforated and partially outdated. Harsh climate, long distances, scarce population and limited economic activity complicate the challenges of building and maintaining the much needed infrastructure. Simultaneously, the provision of public and private services, tourism, commerce, agriculture, manufacturing, logistics and all other livelihoods increasingly depend on digital information processing. The state of affairs adds demand for improved connections and pressure on existing infrastructure.

In addition to infrastructural challenges, people's awareness of and skills to operate safely in the digital environment require supplementation. The development pace of technology and its accelerating application in everyday practices tend to leave people behind. They are expected to adapt to the omnipresence of ICTs which requires additional education, training and support. This development highlights some of the existing socio-cultural-economic divides, while diluting others and creating new ones. Yet, technology ought to be developed to serve the needs, interests and wants of people and communities living in the region, for solving everyday practical problems and improving the quality of life. For this reason, people and communities should have a say in digitalisation. Inclusiveness and empowerment are equally important cornerstones of human security alongside the reduction of fear, want and indignity.

3 Assessment

The following assessment of digital security in the Barents Region examines six themes:[10]

1 access to digital networks
2 availability of digital services
3 access to (relevant) information
4 digital awareness and skills
5 digital inclusion policies and protection of human rights
6 state of cybercrime and digital abuse.

The aforementioned themes are interlinked. Moreover, they directly and/or indirectly affect all recognised dimensions of human security from economic to political security. They have the ability to either enhance or mitigate threats to human security, as well as to either empower or disable individuals and communities to act on behalf of their own wellbeing. As there is no comprehensive regional data available on the aforementioned themes, aggregated national data is utilised as an indicative reference point.

The wellbeing of individuals and communities is not the main concern in national digitalisation and information/cybersecurity policies. Instead, their primary reasoning lies in economic development, efficiency gains and/or smooth and secured functioning of society. However, the policies promise increased opportunities and ease of life to individuals, while allocating them a share of responsibility for security production. In information/cybersecurity the human being is often named as the weakest link of the security chain or as a threat (e.g. Singer and Friedman 2014, 64–65, 77–103; cf. Carr 2012, 147; Dunn Cavelty 2014, 703–704), whereas digital security as an aspect of human security concentrates on the human being as its main referent object and an active producer of security. Regardless of the perspective, improving people's digital awareness and skills is the main policy recommendation. Yet improved awareness and skills are of little help if inadequate or lacking access to digital networks constitutes an obstacle to utilising the opportunities provided by digitalisation.

3.1 Access to digital networks

Relevant and up-to-date[11] access data concerning the Barents Region has been collected from ITU Country profiles in 2015 and collated into Table 2.10.1. The number of fixed or mobile broadband subscriptions is higher in the Nordic countries than in Russia, which directly impacts the percentage of population using internet. This chapter does not aim to map national information infrastructure, compare business-to-business or consumer

Table 2.10.1 Access to digital networks according to ITU country profiles in 2015[12]

Country location	*Finland*	*Norway*	*Russia*	*Sweden*
Fixed telephone subscriptions per 100 inhabitants	9.8	21.2**	25.7	36.7
Mobile telephone subscriptions per 100 inhabitants	135.5	116.1**	160	130.4
Fixed broadband subscriptions per 100 inhabitants	32.3	38.8**	17.5	34.1
Mobile broadband subscriptions per 100 inhabitants	144	88.8**	71.3	122.1
Households with a computer (%)	89.3	95.4**	72.5	88.2
Households with internet access at home (%)	89.9	93.1**	72	91
Individuals using internet (%)	92.6	96.3**	73.4	90.6
Definition of universal service/ access exists	Y	Y**	N*/Y***	Y**
Universal service/access policy adopted	Y	Y**	Y***	Y**
Operators under universal service/access obligation	Designated universal service provider	Incumbent fixed line operator(s)**	State-owned Rostelecom	N (obligations allocated on a competitive basis)**
Means of financing operators obligation	So far no subsidies (the option remains)	Cross-subsidy between own services (internal financing)**	Universal service fund	N**

Y = yes, N = no
* data available from 1999
** data available from 2014
*** In 2014, RIA and Comnews reported on announced state operator's investment programme: https://ria.ru/economy/20140220/995994372.html and www.comnews.ru/node/80789 [accessed 2.2.2017]

prices of equipment and subscriptions, or investigate regulatory frameworks or market structures in the Barents Region. Therefore, it also abstains from speculating the causes of national differences but settles for describing the current state of affairs.

An exception to the aforementioned rule is universal service/access policies. These policies constitute a state intervention in the market which has turned out to be necessary for the provision of network access and digital services in the Barents Region (e.g. Lapin liitto 2013, 27–28, 41; Norrbotten 2014, 11, 20–23). Aggregated national data, such as ITU data, does not give insight into the substantial regional differences in network access. Universal service/access policies have been the means for national, regional and local administrations to advance digital network development in areas characterised by challenging geographical features, long distances, harsh climates, sparse and ageing populations, disadvantageous socio-economic situations, as well as a multiplicity of cultural traditions and spoken languages.

Universal service/access policies exist across the Barents Region. The means of organising and funding the provision of universal service/access varies between the countries. Yet the basic idea remains the same: to provide digital opportunities for authorities, businesses, other organisations and individuals regardless of their physical location within the state. Consequently, digitisation enables the provision of services in a novel format, which has a number of implications.

First, in principle, services can be provided on a transnational market to a bigger pool of clients. Simultaneously, physical services can be withdrawn from rural areas while reasoning this development with the provision of digital substitutes. Second, full participation in digitalising societies requires access to the digital environment and adequate skills and knowledge to operate in it. Third, both physical and logical layers of cyberspace[13] need to be successfully protected so that everyday life continues smoothly. Fourth, the social layer of cyberspace[14] needs its security arrangements to be upheld so that people's, businesses' and other organisations' trust in the digital environment is maintained. Finally, utilisation of the opportunities provided by digitisation should not endanger the realisation of human rights in any situation. What the latter entails is currently under intensive political, legal and ethical discussion.

The extent and quality of information infrastructure in the Barents Region is an acknowledged, continuing challenge (Lapin liitto 2013; Norrbotten 2014; AEC 2016). As technologies develop, and digital services require more bandwidth, it is no longer enough, for instance, that a digital connection simply exists; it also needs to be good enough to run and use digital services. Therefore, not only end devices but also network infrastructures require constant updating.[15] The maintenance, update and extension of information infrastructure generate high costs in the region due to its aforementioned characteristics and scarcity of skilled people (AEC 2016, 12–13). Current reports note problems, for instance, in running errands digitally with government authorities or businesses, accepting card/mobile payments and utilising ICTs in agriculture (Norrbotten 2014, 15–16, 25). In general, inadequate network infrastructure hinders all economic activities and development in the region (cf. Delaunay 2014, 5). However, the establishment of, for instance, novel extractive industrial sites may bring digital development with them (AEC 2016, 19, 28). The Arctic Economic Council (2016, 22–26) has also listed a number of planned or on-going transnational infrastructure development projects such as NxtVn's Oulu Nordic Express, the Russian Optical Trans-Arctic Submarine Cable System and Svalbard Undersea Cable System.

3.2 Availability of digital services

The development of digital services in both private and public sectors is a continuous undertaking. Regardless of the fast technological development, digitalisation proceeds only gradually and requires adaptation in both sectors, which leads to structural adjustments, novel operating models and/or 'new ways of doing old things'. For example, health and social services have been digitised to a great extent in Lapland.[16] These technology-mediated services are designed for the use of customers, as well as of social and health care professionals. Customer services include (pre-filled) online forms, online appointment and advice, as well as counters and metrics for self-evaluation and self-tracking. In addition, real-time connection with technical support can be arranged to the customer's home or to a physical service point[17] for an appointment or a negotiation. Services for professionals consist of online consultation (also a real-time possibility) and a tool kit to support one's work. The services are developed in a particular project funded through national and EU funding schemes.[18]

Adjustments and renewals often generate opposition and it takes time for people and organisations to pick up digital services. However, the current societal trajectory creates a situation in which people may have no alternatives (Kilpeläinen 2016, 61). Yet digitisation occasionally fails, making services difficult to use or unaccountable. The failure leads to new rounds of adjustments, redesigning and renewals, which all increase digitisation costs. One important difference in service digitisation in private and public sectors is that due to competition, the former needs to be sensitive to customer needs and preferences, whereas – and regardless of national policy texts – development in the latter is often driven by intra-organisational requirements (e.g. Lapin liitto 2013, 21).

More and more digital services are becoming available in the Barents Region to serve the needs and interests of businesses operating in the area, tourism industry, the authorities, as well as local people and communities. Nonetheless, gaps exist in recognising stakeholder needs and developing digital solutions to them. Reasons for the state of affairs may stem, for instance, from inadequate research; lacking skills, interest or funding; lacking partners within the technology industry; disputes over data ownership; and lacking imagination. Moreover, the prospective client base and hence incoming cash flow for businesses is restricted in services tailored to the needs of the Barents

Table 2.10.2 Users of digital public services in the Nordic countries in the Barents Region as in 2015

Country users of digital public services	*Finland*	*Norway*	*Sweden*	*EU average*
eGovernment (% of individuals, out of internet users aged 16–79)	63	59	49	31
Pre-filled forms (score 0–100)	87	79	75	49
Online service completion (score 0–100)	93	90	89	81

region. Developing services across national borders within the region would already increase the pool of potential clients. However, it could also increase costs, for example, because services would need to be provided in a number of languages. Lack of language versions, as well as problems in scalability to different end devices, is a problem recognised in current digital service provision. For example, in Finland it is difficult to run errands digitally in any of the Sámi languages or in Kven language in Norway.

European Union's DESI reports[19] for the Nordic countries in 2016 provide data on the adoption of digital public services. The data has been collected into Table 2.10.2. Unfortunately, the EU does not give a similar estimation for Russia.

In general, the Nordic countries run clearly ahead of the EU average in the development and adoption of digital services. For example, around half of the internet-users aged between 16 and 79 in Finland, Norway and Sweden have used e-government services. Pre-filled forms and online service completion are familiar to the people as well. Yet both technology and shareholder needs evolve constantly, which calls for further service development.

3.3 Access to (relevant) information

The question of access to (relevant) information is sub-divided into two parts in this chapter. The first part discusses the availability of culturally and linguistically relevant information, information related to livelihoods within the region, as well as gaps in research. The second part concentrates on the freedom of information in the Barents Region.

3.3.1 Access to culturally, linguistically and economically relevant information

Digitalisation has multiplied the amount of online information so that there is a vivid discussion on 'information overload'. For instance, EMC2/IDC estimated in 2014 that the size of the 'digital universe' will reach 44 zettabytes by 2020. However, excess of information does not signify that all relevant information would be digitally accessible. As noted earlier, existing digital services, as well as administrative or business-related information, either are not or are only partially available in minority languages in the Barents Region (*see also*

Larsen and Fondahl 2014, 118). From the perspective of human security, this state of affairs constitutes a threat to people's everyday wellbeing in digitalising societies. Moreover, what is often referred to as 'indigenous knowledge[20]' is only gradually finding its way to the digital environment due to, for example, constricted interest in digitising it, inadequate value given to it (e.g. Artic Council 2016, xi), and individuals' and communities' unwillingness to share it. Whether all information should be shared through digital networks is a question beyond the scope of this chapter.

Substantial research needs be carried out on the ways that individuals and communities in the Barents Region are using, cannot use, would like to use and/or would refrain from using ICTs and digital information, as well as for which purposes. Research also ought to map the existing lacks and gaps in relevant information provision. For instance, economies in the region are characterised as much by small-scale traditional production as by large-scale production and natural resource distribution to international markets (Larsen and Fondahl 2014, 151). Thus, whether and in which ways digitalisation may improve or hinder economic development needs to be studied on a case-by-case basis instead of making generalised assumptions, like the existing national and regional strategies tend to do. Finding out what kind of data would help best and how it can be analysed and utilised should be the driver in developing innovative solutions. People and communities living in the region should be included in the group of stakeholders discussing preferred directions of development. Similarly, research ought to cover the ways of cultural identity formation in the digital environment as a counterbalance to the presumed threats to cultural identity established by digitalisation (e.g. Larsen and Fondahl 2014, 128).

3.3.2 Freedom of information

According to UNHRC (2011, 1) states are increasingly censoring information online through:

> arbitrary blocking or filtering of content; criminalization of legitimate expression; imposition of intermediary liability; disconnecting users from Internet access, including on the basis of intellectual property rights law; cyberattacks; and inadequate protection of the right to privacy and data protection.

While there are certain types of data to which access may be restricted, including child pornography; hate speech; defamation; direct and public incitement to commit genocide; and advocacy of national, racial or religious hatred that constitutes incitement to discrimination, hostility or violence (UNHRC 2011, 8), the overarching principle should be that of information freedom.

Freedom House[21] publishes freedom indices of which 'Freedom in the World 2017' gives Finland, Norway and Sweden full a 100 points out of 25 political freedom and civil liberties indicators with the labelling of 'free'. Russia receives 20 points with the labelling of 'not free'. The institute also publishes a 'Freedom on the Net' report which, unfortunately, does not assess the Nordic countries. In the 2016 version of the report, Russia's information space was estimated as 'not free' (Freedom House 2017, 20–24). The country report pointed out violations in user rights, imprisonment of social media users, content blocking and filtering, as well as violent or cyberattacks on online activists.[22] There is no data available specific to the Barents Region, but similar imbalance in freedom of information can be expected to exist regionally.

3.4 Digital awareness and skills

Access to digital networks does not guarantee that people and communities will adopt ICTs and digital forms of service provision. Moreover, it does not ensure that they are aware of and able to estimate risks inherent in the use of ICTs or know how to operate in the digital environment in a safe and secure manner.

European Union's DESI reports for the Nordic countries in 2016 provide statistics about individuals having basic digital skills, being ICT specialists, or STEM[23] graduates. The data has been collected and collated into Table 2.10.3. Unfortunately, the EU does not give a similar estimation for Russia. Finland has the highest percentage of workforce in the ICT sector within the entire the EU, while Sweden has the second highest. However, demand for highly skilled ICT professionals remains high throughout the Barents Region (e.g. HS 17.5.2017).

According to aggregated Nordic data[24] provided by FireEye (2016, 6) 35 per cent of the consumers in the Nordic countries believe they have 'full awareness and understanding' of the potential impact of data breaches on them personally. The respective percentages for having 'partial awareness and understanding' and 'little to no awareness and understanding' are 52 and 12. Furthermore, only 39 per cent of the respondents thought security as 'an important or main consideration' when purchasing products or services. Yet, 45 per cent would 'consider paying more to work with a provider with better data security' (FireEye 2016, 15–16). The data does not provide any information on national – or regional – differences and the sample is very small.

Table 2.10.3 Digital skill levels in the Nordic countries in the Barents Region

Country skills level	*Finland*	*Norway*	*Sweden*	*EU average*
Basic digital skills (% of individuals aged 16–74)	75*	80*	72*	55*
ICT professionals (% of employed individuals)	6.7**	4.8**	6.0**	3.7**
STEM graduates (per 1,000 individuals aged 20–29)	22***	13***	15***	18***

* data available from 2015
** data available from 2014
*** Data available from 2013

However, it does support the general feeling amongst cybersecurity experts that people are generally unaware of the risks inherent in the use of ICTs, even if the level of awareness has been rising due to, for example, increased public media attention given to digital security. Next to that, people tend to bypass security concerns if the usability of digital products or services is at the other end of the scale, due to peer pressure and/or if access to relevant digital content through other channels is unavailable.

The burden of responsibility for improving one's digital awareness and skills has largely been left to individuals. Technology development has been so fast that the majority of people in work life have not received adequate training for the safe and secure use of ICTs at school, in later studies, or through employer provided training. Skills development has depended on their own interest in technology and its use. National policies in the Barents Region alleviate part of the problem by increasing education on digital skills at schools and other academic institutions. Employers are gradually waking up to the problem, and individuals can improve their know-how through training provided by NGOs or adult education institutes, among others. At best, public media plays an important role in raising digital awareness, whereas social media continues to test people's media literacy. Media convergence and the pace of dissemination in the digital environment complicates the challenge of differentiating correct information from incorrect.

3.5 Digital inclusion policies and protection of human rights

Digital inclusion as a question of access to digital networks has been addressed above. Nonetheless, it inheres further questions related to people's socio-economic standing, gender, minority position, disability, age, and attitude (e.g. UNHRC 2011, 17; Cruz-Jesus et al. 2012). In the Barents Region, people and organisations are buying or leasing their own ICTs and network subscriptions. The chances to acquire equipment and connections depend directly on one's socio-economic standing. In addition, the willingness to 'be connected' stems from one's own attitudes and opinions. For example, in Norrbotten it is estimated that around 30 000 inhabitants reside outside the 'digital world' (Norrbotten 2014, 17). In Finland, internet access has been perceived as a fundamental right which, again, has led to the question of whether it should be included in publicly provided social security. Currently, public internet access is already provided, for example, in libraries, town halls or community houses.

Policies for removing the obstacles to access and utilisation of the digital environment exist in all Nordic countries. These cover the questions of age and age-related additional challenges such as eyesight and hearing, as well as the ease of use of end devices and digital services. They also address challenges related to people with disabilities. Means of addressing such challenges include, for instance, requirements for being able to adjust the font size on websites, options to listen to the text on websites, as well as different technical aids and support for using ICTs. Inequality questions related to gender and minority positions are tackled with, for example, language versions and training. However, a lot remains to be done before the digital environment is equally friendly to all its utilisers.

In 2011, a report of the Special Rapporteur on the promotion and protection of the right to freedom of opinion and expression to the UN Human Rights Council highlighted the importance of the digital environment 'not only to enable individuals to exercise their right to freedom of opinion and expression, but also a range of other human rights' (UNHRC 2011, 1). The former was declared as 'a fundamental right on its own accord', as well as 'an 'enabler' of other rights', including a number of economic, social and cultural rights,[25] as well as civil and political rights[26] (UNHRC 2011, 7). Moreover, '[t]he right to privacy is essential for individuals to express themselves freely' (UNHRC 2011, 7, 15). Thus,

> every individual should have the right to ascertain in an intelligible form, whether, and if so, what personal data is stored in automatic data files, and for what purposes. Every individual should also be able to ascertain which public authorities or private individuals or bodies control or may control their files.
>
> (UNHRC 2011, 16)

The general principles in and state responsibility for protecting human rights apply as much in the digital environment as in the physical environment. Furthermore, the contemporary merging of digital and physical environments highlights the impossibility of keeping the two apart. The current tendency throughout the Barents Region to curtail privacy and to strive for exemptions in human rights protection in the name of national security is an alarming development. Renewals of intelligence legislation are currently being prepared in Finland and Sweden, whereas Russia already has an efficient online surveillance and data collection system in place.[27] Simultaneously, many NGOs aim to bring forward and keep in discussion the concern over human rights protection in the digital environment.

3.6 State of cybercrime and digital abuse

The Council of Europe's Convention on Cybercrime, the so-called Budapest Convention,[28] divides cybercrime into four categories: (1) offences against the confidentiality, integrity and availability of computer data and systems, (2) computer-related offences, (3) content-related offences, and (4) offences related to infringements of copyright and related rights (European Council 2001). The first category includes actions such as illegal access, illegal interception, data interference, system interference and misuse of devices. The second category consists of fraud and forgery, whereas the third deals with child pornography. The convention recommends states to legislate cybercrime and develop mutual cooperation in countering it. More simply, cybercrime can be divided, on the one hand, to crimes targeting ICTs and digital networks and, on the other hand, to crimes utilising ICTs and digital networks.[29]

Nobody has a full situational picture of cybercrime. A large number of crimes go unreported for reasons such as not being detected or not being perceived as worth reporting, uncertainties about what cybercrime is and where to report it, as well as pressures to uphold organisational reputation (e.g. Helsingin seudun kauppakamari 2015, 1, 7; Leppänen et al. 2016, 20–21). Moreover, trust in the ability of the police to solve cybercrimes has been contested for reasons such as a late start in systematic collection and analysis of cybercrime data (e.g. Leppänen et al. 2016, 15–17, 24; Clough 2010, 7–8, 15). According to Nasjonal sikkerhetsmyndighet,[30] network operations and digital espionage are the gravest cyber threat to the state, critical infrastructure owners and/or operators, entities responsible for functions vital to society, and high-technology companies in Norway. However, the rest of companies should mainly worry about (economically motivated) cybercrime (NSM 2016, 5, 31) In Sweden, Myndigheten för samhällsskydd och beredskap[31] estimated in 2015 that ICT-related organised crime is most often financially motivated and that the crime-as-service[32] sector is growing. Malware-as-service,[33] ransomware[34] and identity thefts are also increasing. Mixing of traditional and network-based crime is becoming more common and complex thus requiring intensified international cooperation in countering it (MSB 2015, 35–39). In general, cybercrime is well organised, globally connected and technically advanced. Yet people and organisations continue to fall victim to simple scams and hand over sensitive information carelessly when just being asked.

In its annual report for 2016 on information security, Viestintävirasto[35] in Finland lists internet scams and phishing[36] as flourishing businesses; ransomware as having dominated the year of malware; network espionage and advanced persistent threats[37] as headaches for an increasing number of organisations; and (distributed) denial-of-service attacks[38] as having increased in volume and become a tool for blackmail. However, defective equipment is still the main cause for network disturbances. In addition, advancement of Internet of Things[39] is beginning to cause novel and reformed cybersecurity concerns (Viestintävirasto 2017, 5–19). According to FireEye (2015, 3–4, 6, 11), regional characteristics such as rich natural resources, innovations in healthcare and renewable energy, high levels of connectedness and high-technology industry, strong shipping industry, as well as transparent governance make the Nordic states targets for malicious digital activity.

The state of cybercrime in Russia is very different from the Nordic countries. According to Kaspersky Lab (2015, 2) '[t]he Russian-language cybercrime market is known all over the world' due to frequent global media coverage and the accessibility of online platforms used by the criminal community. In 2011, a consolidation within the cybercrime market was noted: major cybercrime groups were emerging to overtake the previously disorganised market. Simultaneously, traditional organised crime groups were penetrating the cybercrime market. This development increased both the value of Russian cybercrime and its sophistication (Group-IB 2012, 7–8). In addition, the relationship between cybercrime groups and the Russian administration has been under speculation for years (e.g. FireEye 2014).

In 2015, all forms of financially motivated cybercrime were prevalent in Russia (targeting organisations both inside and outside the state borders). Products and services were sold on the cybercrime market in different combinations to enable mainly five types of crime: distributed denial-of-service attacks, theft of personal data or data to access virtual money, money thefts from organisations' accounts, domestic or corporate espionage, and blocking access to data for the purpose of extortion. For Russian cybercriminals the risk of prosecution remains low. Cybercrime stays attractive also because of the lack of established mechanisms for international cooperation (Kaspersky Lab 2015, 5–6, 21). Moreover, Russian cybercrime is becoming increasingly automatised (Trend Micro 2016, 6).

In the Barents Region, financially motivated cybercrime is likely to be strongly related to existing industries and digitising livelihoods: tourism, high-technology companies, extractive industries, retail and logistics are likely targets of cybercrime. Both the providers in and the clients of these industries suffer the consequences of successful criminal activities, for example, in the form of financial and reputational losses, as well as delivery delays and/or outages. Enduring problems with organisational continuity planning and risk management turn into questions of human security when people and communities begin to experience the consequences in their everyday life. While individuals can also fall victim to financial cybercrime, hate speech (particularly in social media), cyber bullying and digital abuse (verbal or visual) are highly likely phenomena as well. The aforementioned forms of digital abuse still often go unnoticed; are likely to relate to minority status, disabilities, sexual orientation, gender or whichever trait another person finds 'different'; and should be better investigated.

4 Conclusions

As this chapter has demonstrated, digital security is an increasingly important, multifaceted, omnipresent and largely unexamined aspect of human security in the Barents Region. It affects the everyday life of people and communities residing in the region both directly, for example, in the form of access to and safety of digital services or potential for digital abuse and human rights violations, and indirectly, for instance, as a consequence of cybercrime against businesses operating in the region.

Human security issues related to digitalisation have usually not been addressed in the information/cybersecurity agenda, but within the frameworks of universal service/access and national broadband policies, information society or digitalisation policies, or as questions of socio-economic (in)equality. However, this chapter argues that these issues need to be brought onto the security agenda as questions of human security in the digital environment. Only thus will digital awareness and skills, inclusivity of the digital environment, ICT-user safety and security, as well as access to digital networks, services and relevant information receive the attention they deserve. For the ease of people's and communities' everyday life, these themes are at least as important as questions of data security and critical infrastructure protection.

Digital security is not a constant across the Barents Region but varies, for instance, on the basis of national regulation and policies, market structures, the state of regional development and inhabitants' socio-economic standing, majority/minority point of view and the accessibility of digital services. In order to provide more detailed information and access to all stakeholders to support the regional digital development serving local needs and interests, substantial research has to be carried out. This research should concentrate on the ways that individuals and communities in the Barents Region are using, cannot use, would like to use and/or would refrain from using ICTs and digital information, as well as for which purposes. Whether, and in which ways, digitalisation may improve or hinder regional development needs to be studied on a case-by-case basis instead of making generalised assumptions. Finding out what kind of data would help best and how it can be analysed should be the driver in developing innovative solutions.

In addition to research that would support better decision-making in both public and private sectors, the implementation of existing digitalisation and information/cybersecurity strategies, policies and programmes needs to be improved. Examining the applicability of existing regulation that supports human security needs to be intensified in the digital environment. The principle of 'what is unacceptable in the physical environment is also so in the digital environment' ought to be followed more strictly, and in a demonstrable manner. This requires commitment, sufficient resource allocation, general enhancement of digital awareness and skills, willingness and firmness in decision-making, adjustments in or novel regulation when deemed necessary, as well as enhanced cooperation both within the Barents Region and across its borders. Demonstrated decisiveness in tackling digital security concerns as an aspect of human security would, hopefully, encourage similar development outside the region. Moreover, implementation but also reformation of the existing policies to better support human security needs to be opened to all stakeholders in the Barents Region so that people's and communities' needs and fears are addressed in a comprehensive manner. Improvements to inclusion could be achieved through, for example, consultation and integration in different stages of decision-making, as well as active enhancement of public dialogue in digital security.

Notes

1 E.g. United Nations Office for Disarmament Affairs (UNODA) utilises the concept of 'information security', whereas ITU uses 'cybersecurity'.

2 UNDP: United Nations Development Programme

3 National 'information security' strategies and frameworks have existed in the Nordic countries and in Russia since the 1990s. Norway published her cyber security strategy in 2012; Finland in 2013; and Sweden in 2017. Russia has addressed digital security questions under the umbrella of 'international information security' in 2013; in early 2014, a national cyber security strategy was also reported to be in the making (http://tass.com/russia/714091 [30.1.2017]).

4 the UN specialised agency for information and communication technologies (ICTs). See www.itu.int/en/Pages/default.aspx [18.5.2017]

5 Organisation and user's assets include connected computing devices, personnel, infrastructure, applications, services, telecommunications systems, and the totality of transmitted and/or stored information in the cyber environment.

6 International Telecommunication Union (N/D). *Definition of cybersecurity.* http://itu.int/en/ITU-T/studygroups/com17/Pages/cybersecurity.aspx.

7 SANS Institute is a recognised American private company specialising in technical training in information and cybersecurity.

8 SANS Institute (N/D). Information Security Resources. https://sans.org/information-security/.

9 and, thus, avoiding the national security and critical infrastructure focus of 'cybersecurity' and the general information or technology focus of 'information security'

10 Several metrics for assessing national cybersecurity or digitalisation exist, for instance, ITU's Global Cybersecurity Index (GSI) and EU's Digital Economy and Society Index (DESI). Nonetheless, they do not approach digital security from the human security perspective. As being only partially suitable for this chapter, some of their insights have been integrated in the seven themes. UN Human Development Report utilise two digital indicators beneficial to this chapter, both under the heading of 'mobility and communication' and namely 'mobile phone subscriptions (per 100 people)' and 'internet users (% of population)'.

11 Up-to-date meaning that the data has been provided either for 2014 (Sweden; Norway and Russia partially) or 2015 (Finland; Norway and Russia partially).

12 Available through www.itu.int/net4/itu-d/icteye/CountryProfile.aspx [9.12.2016] (excluding data about Russia in 2014).

13 Physical layer: the physical location of network elements, as well as all hardware and infrastructure (wired, wireless, and optical) that supports the network and its physical connectors (wires, cables, radio frequency, routers, servers, and computers). Logical layer: the logical network component which is technical in nature and consists of the logical connections that exist between network nodes (i.e. any device connected to the network). www.acqnotes.com/acqnote/careerfields/cyberspace [1.2.2017]

14 Social layer: a person's identification or persona on the network (e.g. e-mail address, computer IP address or cell phone number) and the people actually on the network. www.acqnotes.com/acqnote/careerfields/cyberspace [1.2.2017]

15 AEC (2016, 17–18) estimates that also in the future the information infrastructure in the Barents Region will consist of a compilation of different network technologies.

16 See www.sosiaalikollega.fi/virtu.fi/ [5.2.2017]. Available only in Finnish.

17 Service points consist of all necessary equipment, software and connections for using the digital services. They are provided by municipalities and located in public spaces such as libraries, town halls or community houses.

18 Virtu.fi-presentation. PowerPoint-presentation explaining the project and introducing the services. Acquired through email from the project leader, Pirttijärvi, M. (the Centre of Excellence on Social Welfare in Northern Finland), in 17.8.2016. In the possession of the author.

19 European Commission. 2016. The Digital Economy and Society Index (DESI). Progress by country reports are available through https://ec.europa.eu/digital-single-market/en/progress-country.
20 Here understood as local knowledge unique to a particular culture or community
21 Freedom House is an independent American watchdog organisation 'dedicated to the expansion of freedom and democracy around the world' and operating through 'a combination of analysis, advocacy, and action' (https://freedomhouse.org/about-us [14.2.2017]).
22 Freedom House's 'Freedom on the Net' 2016 country report on Russia. https://freedomhouse.org/sites/default/files/FOTN%202016%20Russia.pdf; see also Kshetri (2016, 211–221).
23 STEM: Science, Technology, Engineering, and Mathematics
24 Countries included are Denmark, Finland, Iceland, Norway and Sweden. A market research company interviewed 500 selected consumers aged 18 or over online in April 2016 for the report (FireEye 2016, 2.)
25 such as the right to education and the right to take part in cultural life and to enjoy the benefits of scientific progress and its applications (UNHRC 2011, 7)
26 such as the rights to freedom of association and assembly (ibid.)
27 SORM as a 'system for operational investigative measures'. In 1996, the first version of the system was established to monitor telephone conversations. In 1998, it was extended to internet. In practice, 'Internet Service Providers (ISPs) must install a special device on their servers to allow the FSB [Federal Security Service] to track all credit card transactions, e-mail messages and web use.' http://resources.infosecinstitute.com/ russia-controls-internet/ [1.7.2014/14.2.2017] In 2014, a new wiretapping system requirements were introduced: https://rublacklist.net/8827/ [14.2.2017] See also Freedom House's 'Freedom on the Net' country report on Russia.
28 The convention was signed in Budapest on Nov 23, 2001. It has become the first de facto international treaty on cybercrime.
29 Poliisi (n.d.). Kyberrikollisuus: https://poliisi.fi/rikokset/kyberrikollisuus.
30 Norwegian National Security Authority
31 Swedish Civil Contingencies Agency
32 Crime-as-service: illegal business model in which an entire cybercrime or part(s) of it is provided as a service in the black market.
33 Malware-as-service: ready-made or custom-made malware, build-it-yourself kits and hosted management services for deploying malware that can be bought as a service in the black market.
34 Ransomware: malware which after infecting a computer encrypts information on it so that the information becomes unavailable to its legitimate user and a call for ransom can be made. The criminal claims to decrypt the information when ransom has been paid.
35 Finnish Communications Regulatory Authority (hosts Cyber Security Centre)
36 Phishing: attempts to acquire sensitive information like credit card numbers, social security codes, passwords and usernames for misuse. Carried out by disguising as a trustworthy organisation or a representative of one in digital communication.
37 Advanced persistent threat: a series of sophisticated cyber-attacks carried out in a stealthy manner so that the selected target should not become aware of the on-going operation.
38 Denial-of-service attack: an attempt to make a digital resource unavailable to its legitimate users by disrupting services e.g. by overloading the resource with service requests.
39 Internet of Things: development in which more and more equipment and appliances are connected to internet and communicate without human interference.

References

Arctic Council. 2016. *Arctic Resilience Report*. https://sei-international.org/mediamanager/documents/Publications/ArcticResilienceReport-2016.pdf.

Arctic Economic Council (AEC). Telecommunications Infrastructure Working Group. 2016. *Arctic Broadband. Recommendations for an Interconnected Arctic*. http://arcticeconomiccouncil.com/wp-content/uploads/2017/01/AEC-Arctic-Broadband.pdf.

Carr, J. 2012. *Inside Cyber Warfare*. Sebastopol (CA): O'Reilly Media.

Clough, J. 2010. *Principles of Cybercrime*. Cambridge: Cambridge University Press.

Cruz-Jesus, F., Oliveira, T. and Bacao, F. 2012. Digital Divide Across the European Union. *Information and Management* 49(6): 278–291.

Delaunay, M. 2014. The Arctic: A New Internet Highway? Policy Brief. *Arctic Yearbook*. http://arcticyearbook.com/images/Arcticles_2014/BN/Delaunay_AY_2014_FINAL.pdf.

Dunn Cavelty, M. 2014. Breaking the Cyber-Security Dilemma: Aligning Security Needs and Removing Vulnerabilities. *Science and Engineering Ethics* 20(3): 701–715.

EMC2/IDC. 2014. The Digital Universe of Opportunities: Rich Data and the Increasing Value of the Internet of Things. Interactive report. https://emc.com/leadership/digital-universe/2014iview/index.htm.

European Council. 2001. *Convention on Cybercrime*. 23.11.2004, Budapest. European Treaty Series No.185. http://europarl.europa.eu/meetdocs/2014_2019/documents/libe/dv/7_conv_budapest_/7_conv_budapest_en.pdf.

FireEye. 2014. APT28: A Window into Russia's Cyber Espionage Operations?https://fireeye.com/content/dam/fireeye-www/global/en/current-threats/pdfs/rpt-apt28.pdf.

FireEye. 2015. Cyber Threats to the Nordic Region. Threat Intelligence report. https://fireeye.com/content/dam/fireeye-www/global/en/current-threats/pdfs/rpt-nordic-threat-landscape.pdf.

FireEye. 2016. Beyond the Bottom Line: The Real Cost of Data Breaches. www2.fireeye.com/WEB-Real-Cost-of-Data-Breaches.html.

Freedom House. 2017. Populists and Autocrats: The Dual Threat to Global Democracy. Freedom in the World 2017. https://freedomhouse.org/sites/default/files/FH_FIW_2017_Report_Final.pdf.

Group-IB. 2012. State and Trends of the 'Russian' Digital Crime Market 2011. http://group-ib.com/images/media/Group-IB_Report_2011_ENG.pdf.

Helsingin Sanomat. 2017. Töitä olisi heti 9 000 osaavalle koodarille, mutta tekijöitä ei löydy – Vaikka maksaisit mitä, se ei muuttaisi tilannetta, sanoo ohjelmistoyrityksen teknologiajohtaja. Kari Räisänen. http://hs.fi/talous/art-2000005225501.html.

Helsingin seudun kauppakamari. 2015. Yrityksiin kohdistuvat kyberuhat 2015. http://helsinki.chamber.fi/media/filer_public/36/0f/360fddcd-4cfe-41a6-ab89-c028aa0bf15c/kyberturvallisuus_2015.pdf.

International Institute for Strategic Studies (IISS). 2015. *Evolution of the Cyber Domain: The Implications for National and Global Security*. An IISS Strategy Dossier. London: IISS.

Kaspersky Lab. 2015. Russian Financial Cybercrime: How it Works. https://securelist.com/files/2015/11/Kaspersky_Lab_cybercrime_underground_report_eng_v1_0.pdf.

Kilpeläinen, A. 2016. Teknologiavälitteisyys kyläläisten arjessa. Tutkimus ikääntyvien sivukylien teknologiavälitteisyydestä ja sen rajapinnoista maaseutusosiaalityöhön. Doctoral dissertation, University of Lapland. Rovaniemi: Acta Universitatis Lapponiensis 316.

Kshetri, N. 2016. *The Quest to Cyber Superiority. Cybersecurity Regulations, Frameworks, and Strategies of Major Economies.* Springer.

Kwalwasser, H. 2009. Internet Governance. In *Cyberpower and National Security* edited by F.D. Kramer, S. Starr and L.K. Wentz. Dulles (VA): Potomac Books, 491–524.

Lapin liitto/Regional Council of Lapland. 2013. Lapin digiohjelma 2020. http://lappi.fi/lapinliitto/c/document_library/get_file?folderId=1457612&name=DLFE-21300.pdf.

Larsen, N.L. and Fondahl, G. (eds) 2014. *Arctic Human Development Report (AHDR). Regional Processes and Global Linkages.* Nordic Council of Ministers.

Leppänen, A., Linderborg, K. and SaarimäkiJ. 2016. Tietoverkkorikollisuuden tilannekuva. Valtioneuvoston selvitys- ja tutkimustoiminnan julkaisusarja 17/2016. Valtioneuvoston kanslia. http://vnk.fi/documents/10616/2009122/17_Tietoverkkorikollisuuden+tilannekuva.pdf/6ef911d2-cbe8-43bd-aafa-e10ed573f28a?version=1.0.

Länsstyrelsen Norrbotten. 2014. Digital Agenda. Norrbotten. http://lansstyrelsen.se/norrbotten/SiteCollectionDocuments/Sv/samhallsplanering-och-kulturmiljo/infrastruktur-och-it/Bredband/Digital%20Agenda%20BD%20webbversion.pdf.

Maurer, T. 2011. *Cyber norm emergence at the United Nations – An Analysis of the Activities at the UN Regarding Cyber-security. Discussion Paper #2011–11.* Cambridge (MA): Belfer Center for Science and International Affairs, Harvard Kennedy School. http://un.org/en/ecosoc/cybersecurity/maurer-cyber-norm-dp-2011-11.pdf.

Myndigheten för samhällsskydd och beredskap (MSB). 2015. Informationssäkerhet – trender 2015. https://msb.se/RibData/Filer/pdf/27494.pdf.

Nasjonal sikkerhetsmyndighet (NSM). 2016. Helhetlig IKT-risikobilde 2016. https://nsm.stat.no/globalassets/rapporter/nsm_helhetlig_ikt_risikobilde_2016_web_enkel.pdf

Singer, P.W. and Friedman, A. 2014. *Cybersecurity and Cyberwar. What Everyone Needs to Know.* New York: Oxford University Press.

Trend Micro. 2016. Cybercrime and the Deep Web. https://trendmicro.de/cloud-content/us/pdfs/security-intelligence/white-papers/wp-cybercrime-and-the-deep-web.pdf.

United Nations Development Programme. 1994. *Human Development Report.* http://hdr.undp.org/sites/default/files/reports/255/hdr_1994_en_complete_nostats.pdf.

United Nations General Assembly. 1999. *Developments in the Field of Information and Telecommunications in the Context of International Security.* A/RES/53/70. http://un.org/ga/search/view_doc.asp?symbol=A/RES/53/70.

United Nations Human Rights Council (UNHRC). 2011. *Report of the Special Rapporteur on the Promotion and Protection of the Right to Freedom of Opinion and Expression.* A/HCR/17/27. www2.ohchr.org/English/bodies/hrcouncil/docs/17session/A.HRC.17.27_en.PDF

United Nations Office for Disarmament Affairs. (n.d.). *Developments in the Field of Information and Telecommunications in the Context of International Security.* www.un.org/disarmament/topics/informationsecurity/.

United Nations Office for Disarmament Affairs. 2015. UNODA Fact Sheet on Developments in the Field of Information and Telecommunications in the Context of International Security. https://unoda-web.s3-accelerate.amazonaws.com/wp-content/uploads/2015/07/Information-Security-Fact-Sheet-July2015.pdf.

Viestintävirasto. 2017. Tietoturvan vuosi 2016. *Viestintäviraston julkaisu 001/2017 J.* https://viestintavirasto.fi/attachments/tietoturva/Tietoturvan-vuosi_2016_ViVi_29-11-2017_L.pdf.

Conclusion

Kamrul Hossain and Dorothée Cambou

While there is certainly much more that remains to be discussed regarding human security in the Barents Region, one thing is certain: the protection of the environment remains a major source of insecurity in the region. There are several drivers of environmental insecurity, but climate change and the impact of pollution on the ecosystem are central issues in the region. Even if global actions are required to cope with this issue, environmental security challenges can also be mitigated through regional actions. Two such actions include, for example, expanding the Environmental Hot Spots programme, or reducing black carbon emissions in the region. Considering that climate change is already causing widespread challenges to the Arctic environment, it is also necessary to adopt effective measures for adaptation that could set an example for other countries. In this regard, we argue for the production of a detailed or binding plan to tackle climate change under the auspices of the Barents Euro-Arctic Council. Finally, we suggest that instead of opening new nuclear power facilities, there should be an investment of time, money and expertise in renewable energy sources. Although nuclear energy is sometimes considered an environmentally friendly solution, the safety of the plants and the disposal and management of nuclear waste still require critical attention, as their mismanagement can cause detrimental effects on all aspects of human security.

Concerning economic security, the responsible management of lands and resources requires the foundations for strong regional cooperation to enable economic growth and sustainable development. In our assessment, the need for an attractive environment friendly living condition in order for people to move in, stay or return to the region is emphasised. In order to stimulate economic development and the attractiveness of the region, infrastructural development, such as transportation routes and public service facilities, require improvements. However, tensions surrounding economic and environmental security often arise as a result of conflicting interests. For example, the need to develop industrial and transportation activities to spur economic development conflicts with the need for further environmental protections. In addition, there is also a need to ensure that income generated through economic development is distributed more equally, especially in Russia where poverty is still affecting a large part of the Barents population. Another important dilemma concerns the protection of indigenous groups, whose economic livelihoods are challenged by the adverse impacts of extractive industries. While the rights of indigenous peoples have achieved some progress and recognition in the last decades, conflicts between indigenous peoples and industrial actors such as mining companies are still omnipresent. Therefore, removing the barriers that hinder the engagements of multiple actors, including indigenous peoples, would be beneficial to the economy of the region and would consequently strengthen economic security as well. Ultimately, ensuring that economic development benefits the wellbeing of the population at the local level should be the sacrosanct principle in governing the Barents economy.

Health security is also an issue that is intrinsically connected to the environment and the economy of the region. Living and working conditions in the Barents-Arctic region can pose challenges to both physical and mental health. Given that the health security of the population is also affected by work-life changes associated with climate change and by the increasing industrial activities in the region, a concerted effort by all relevant institutions to fight against the effects of pollution and environmental degradation is necessary. There is also need for intersectional policy actions focusing on disease prevention, safety issues and mental wellbeing. More specifically the spread of HIV and tuberculosis must be contained, and more research on potential new or hidden disease caused by environmental and societal changes are required. Finally, there is also a need to ensure the development of strategies to address the psychological health of Barents peoples and provide support for community and cultural wellbeing, especially for indigenous peoples living in the region.

Concerning the food, water and energy nexus, these three resources are fundamental in guaranteeing the promotion of human wellbeing in the region. In the assessment of human security, we discuss how food, water and energy security are closely interconnected. Because the impacts of climate change and the extraction of resources in the region have significant and negative impacts on the safety of foods, it is important to promote specific and targeted policies that govern food security and safety in the region and which take into account the unique context of the region. To this degree, the importance of an informed and involved community is also underlined in decision-making on factors that impact wellbeing. Moreover, we suggest promotion of policy development in strengthening traditional food supply chains, which would not only improve food security, but also generate income. In this regard, better cross-border collaboration and knowledge sharing amongst stakeholders in the Barents Region is important, requiring the intervention of diverse actors such as governmental authorities, business corporations and local communities.

Regarding water security, integrated policymaking still faces significant gaps. Water governance structures must take into account diverse perspectives from across the region and from different sectors. An interdisciplinary approach to water governance will make decision-making more robust and account for trade-offs and opportunities between resources. Given the lack of existing sources, more research is needed on the impacts of climate change on local freshwater resources in order to understand how communities are adapting water and sanitation systems to the warming and changing environment. There is also great need for more data on the hydrological characteristics of Arctic rivers and watersheds, and for improved methods for the measurement of precipitation. Groundwater assessments and studies should be conducted to determine the quantity and quality of groundwater resources and to establish a baseline. Measureable quantitative indicators of water security should be implemented and monitored in the Arctic regions and more specifically in the Barents area over time.

Consequently, we believe that individual and community interests must be prioritised as the pre-condition for the promotion of food, water and energy security nexus. As discussed, energy security within the Barents Region depends not only on the energy resource in question but also on the interests, aspirations and experiences of the individuals, families and communities that are affected by or dependent on energy-related events, decisions and developments. We also highlight the need for improvements to the reliability and safety of existing energy infrastructures, and for the reduction of dependencies on energy imports. This can be achieved through diversification of energy sources and, where possible, through the decentralisation of energy production. The application of social impact assessments should be improved: one-off pre-project assessments should be replaced by processes of monitoring and managing the social aspects of energy developments throughout the lifespan of the project. Furthermore, efforts should be made to ensure a more balanced representation of consulted communities, and to better understand the impacts of climate change on northern energy production and consumption.

Aside from its resource dimensions, human security in the Barents population also depends on the promotion of security at both individual and community levels, as well as in the political sphere. In relation to personal security, the assessment revealed that current data and research remains insufficient or imminent. Current indicators – from the 2004 and 2014 Arctic Human Development Reports to the Arctic Social Indicators – cannot provide enough data for a comprehensive gender analysis. Nevertheless, these reports recognise gaps and call for gender-disaggregated data. For instance, sex-disaggregated data and gender-sensitive policies are necessary to protect women's rights to personal security, and research and data on gender-based violence remain underdeveloped. Such data would provide a more nuanced understanding of the region's adaptive and transformative capacity while providing more effective policies tailored to the intersectionality of the Barents' population.

Community security is complex, and requires a people-centred approach to determining specific recommendations. In the Barents Region, indigenous groups articulate community security as the preservation of their culture. Specific aspects of cultural preservation can be achieved through the safeguarding and promotion of linguistic and educational rights, as well as land and natural resources rights affecting traditional livelihoods. Our assessment in this regard asserts that cultural preservation and community security require the inclusion of all relevant actors, including indigenous peoples, in decision-making. Indeed, centralising indigenous participation in decision-making processes can result in increased self-determination and social cohesion in addressing immediate and long-term challenges such as climate change, exploitation of natural resources and use of indigenous lands. It is important to highlight strengthened institutional and regulatory frameworks for enhancing the rights of indigenous peoples to self-determination in all Barents countries, as their existence offers cultural significance and cultural diversity for the region. The necessity to protect indigenous rights has become central, in particular to the political security of the region. Yet these recommendations are non-exhaustive and there is still a general lack of data focusing on these issues. Future research is necessary to examine the differences in human security of indigenous peoples in the Barents Region.

More broadly, there is also an urgent need to strengthen the human rights of all citizens, not only of indigenous people. In our view, an obvious problem is the reinforcement of democratic values, human rights and fundamental freedom for all, especially in the Russian Barents. Further, we opine that preserving peace and stability in the region is a condition sine qua non for political security. While military activities in the Barents Region do not at present constitute a threat, militaries and geopolitical events occurring outside of the region have raised significant tensions between states that could adversely affect the regional cooperation with an impact on the region's population. It is therefore necessary to further cooperation both within and outside the region, and strengthen transparency and predictability in the field of military activities.

Digital security is also linked to human security, despite the lack of pre-existing scholarship made from the latter perspective. Therefore, we recognise that omnipresent digitalisation crucially affects human security in the Barents Region, and understanding it is the first step towards ensuring security. Currently, digital security varies across the region, but can be improved through additional contextualised research on its specificities; improved implementation of the existing digitalisation and information/cybersecurity strategies, policies and programmes; and better inclusion of all regional stakeholders in the direction of the digital development. Our assessment calls for the promotion of a comprehensive approach to address the interests, needs and fears related to digitisation. We therefore see the need for improvements in the inclusion of all relevant actors, for instance, through consultation and integration in different stages of decision-making, as well as active enhancement of public dialogue in digital security.

Index

Page numbers in italics refer to figures. Page numbers in bold refer to tables. Page numbers followed by 'n' refer to notes.